The Seven Measures of the World

Piero Martin

Translated from the Italian by Gregory Conti

Yale UNIVERSITY PRESS

New Haven & London

Published with assistance from the foundation
established in memory of Calvin Chapin of the Class of 1788, Yale College.

The translation of this work has been funded by SEPS
Segretariato Europeo per le Pubblicazioni Scientifiche

Via Val d'Aposa 7 – 40123 Bologna – Italy
seps@seps.it — www.seps.it

Originally published in Italy by Editori Laterza as *Le 7 misure del mondo*.

Yale University Press books may be purchased in quantity for educational,
business, or promotional use. For information, please e-mail
sales.press@yale.edu (U.S. office) or sales@yaleup.co.uk (U.K. office).

Set in Janson type by Integrated Publishing Solutions.
Printed in the United States of America.

Library of Congress Control Number: 2022945941
ISBN 978-0-300-26627-6 (hardcover : alk. paper)

A catalogue record for this book is available from the British Library.

This paper meets the requirements of ANSI/NISO Z39.48-1992
(Permanence of Paper).

10 9 8 7 6 5 4 3 2 1

To Paolo, student and mentor

Contents

Contents

The Seven Measures of the World

Introduction

That night, at Große Freiheit 64, Hamburg, the doors of the Indra Musikclub opened at the usual time. It was August 17, 1960, and during the night, the temperature would dip down below 10°C (50°F). Summer was drawing to a close, Elvis Presley was at the top of the charts all over the world with "It's Now or Never," his cover of the Neapolitan standard "O Sole Mio," while in Germany, the French songstress Dalida scored a big hit with a German cover of "Milord," a song recorded a year earlier by Edith Piaf.

For the young women and men waiting outside the Indra on that late summer night, it would have been difficult, well-nigh impossible, to imagine that the virtually unknown band they were about to listen to was on the verge of taking the entire world of popular music on a rollicking decade-long joyride. Equally in the dark about the impact those young rockers were going to have on their company were the top managers of Electric and Musical Industries, otherwise known as EMI.

Founded in London some thirty years earlier, following the merger of the Columbia Graphophone Company and the Gramophone Company, famous for its historic record label "His Master's Voice," EMI was an important player in the music industry. In 1931, one of the company's engineers, Alan Blumlein, had patented the invention of stereophonic recording and reproduction. In the 1960s, EMI produced successful records and created a flourishing research program in the field of electronics. But on August 17, when John Lennon, Paul McCartney, and George Harrison launched into their opening number, things started to change for EMI, too. Harrison, Lennon, and McCartney—together with Pete Best and Stuart Sutcliffe, later to be replaced by Ringo Starr—had just recently become the Beatles, and Hamburg was their first overseas gig. They played at the Indra for forty-eight nights, and then worldwide for another nine years until January 30, 1969, their last live performance on the roof of No. 3 Savile Row, London. What happened in between is legendary.

At the end of World War II, EMI's expertise in electronics was largely in products and instruments used by the military, which it then started marketing to the civilian sector. But the company's economic success came with the rock and pop music explosion of the fifties and sixties. The acquisition of the American company Capitol Records, the success of its artists, and above all the contract it signed in 1962 with the Beatles brought EMI considerable fame and remarkable profits. The projects that EMI engineers were working on in the sixties included the pioneering effort to develop computed tomography in medicine, better known as CT. Today the CT scan is a fundamental instrument in medical diagnostics, making it possible to create high-definition images of the inside of the human body. Its practical application was developed in the EMI laboratories by one of the company's engineers, Godfrey Hounsfield, who applied the theoretical work of the South African physicist Allan MacLeod Cormack.

Hounsfield and Cormack were awarded the Nobel Prize in Medicine in 1979. For years, rumor had it that a major contribution to the discovery of this extraordinary diagnostic tool had been provided by the Fab Four from Liverpool—though this was never claimed by the Beatles themselves. Supposedly, the enormous profits EMI had taken in thanks to their songs had, in part, been used to finance research on CT. In reality, according to an article published by Canadian scientists Zeev Maizlin and Patrick Vos in the *Journal of Computer Assisted Tomography* in 2012, the financial contribution of the British Department of Health and Social Security to the development of the CT scanner was significantly bigger than that of EMI.

Nevertheless, the Beatles' colossal contribution to modern culture is plain for all to see (and hear), as is the fact that medicine, thanks to the laboratories of EMI, now possesses an essential diagnostic tool, which literally helps to save human lives every day. CT is a device that measures the amount of X-radiation emitted by a source and transmitted through the human body and uses that data to reconstruct detailed high-definition images. CT is one of the many examples of how a measurement can provide information about ourselves. This is also the case with measurements of our body temperature, our blood pressure, and the frequency of our heartbeat. To arrive at each of these measurements, we link a number or a set of numbers with a physical quantity, or with the properties of a phenomenon, or with some aspect of nature. This numerical value is obtained by using appropriate tools to compare the physical quantity in question with another physical quantity, known as a unit of measurement. In the case of body temperature, the tool is the thermometer and the unit of measurement is the degree, Celsius or Fahrenheit.

Humans have measured the world from the very beginning. We measure it to learn about it and explore it, to inhabit it, to live together with our fellow humans, to bestow and obtain justice, to relate to divinity. From ancient times on, measuring has interwoven the fabric of human life—just think of the measurement of time and its relationship to our lives, how it shapes our interactions with nature and with the supernatural. Humans measure the world to know our own past, understand the present, and plan the future.

Humans measure using tools that are the fruit of their creativity. In nature, there are many recurring phenomena, like the alternation of day and night or the cycle of the seasons. There are also natural objects whose shape and weight are especially regular, like carob seeds, for example. Thirty thousand years ago, a human living in what is now France engraved an ivory tablet made from a mammoth tooth with what is believed to be a register of the phases of the Moon over the period of a year—a sort of pocket calendar *ante litteram*. It is human ingenuity which used these objects as the inspiration for meridians, scales, and yardsticks. Nature, obviously, works perfectly well even without measurements.

It comes as no surprise, then, that at the dawn of civilization, the first measurements were made by using something that was universally available, something that everyone always has with them: their own bodies. Arms, legs, fingers, and feet are easily accessible tools, and, albeit with a certain variability, they are all more or less of the same size. Five spans of cloth measured by an adult man or woman are about a yard, or a meter, in any part of the world.

Units of measure associated with body parts are found just about everywhere. The cubit, for example, is the distance from the point of the elbow (*cubitus*) to the fingertips, about a foot and a half. It was used by many cultures throughout the Mediterranean basin: Egyptian, Hebrew, Sumerian, Roman, and Greek. The foot is found in China, in ancient Greece, and in Roman culture. In ancient Rome, distance was also measured by the pace, 1,000 of which (*milia passuum*) gave us the Roman mile. And it was also in the Eternal City that Marcus Vitruvius Pollio, who lived from 80 to 20 BCE, wrote *De architectura*, his encyclopedic work dedicated to architecture. In the first chapter of book 3, Vitruvius wrote about symmetry: "The design of a temple is based on symmetry, whose principles must be observed with great care by the architect. They are rooted in proportion." He connected architectural proportion with the proportions of the human body:

> Because the human body is designed by nature so that the face, from the chin to the top of the forehead and to the lowest roots of the hair, is one-tenth of the entire height of the body. . . . The length of the foot is one-sixth of the height of the body; the forearm one-quarter; the width of the chest is also one-quarter. The other parts of the body also have their own symmetrical proportions, and it was by employing them that the famous painters and sculptors of ancient times achieved great and infinite fame.

One of the most famous and iconic drawings by Leonardo da Vinci, *Vitruvian Man*, is conserved in the Accademia Gallery in Venice. Depicting idealized proportions of the body, it owes its name to Vitruvius (even though, it must be added, the gallery specifies on its website that Leonardo was also inspired by Leon Battista Alberti and Euclid). A near contemporary of Leonardo, the German Jacob Köbel, suggested that aligning the feet of sixteen adult men standing outside a church on Sunday morning would produce the unit of measure that he called the *rood*, derived from the analogous German unit *Rute*, which, in turn, comes from the Roman *pertica*, perch or rod.

But humans are also social beings, and measures help us to interact with our fellow humans. As societies gradually expand and become more highly structured, they need common ways of measuring, which, in turn, become powerful cohesive elements for communities. A demand arises for a system of measurement that goes beyond the narrow confines of local villages and towns. It is no coincidence that the great civilizations of the past, such as those in Egypt, Mesopotamia, Greece, and Rome, were all intently focused on the definition of measures. Around 1850 BCE, Pharaoh Senwosret III carefully organized a system of measurement for tillable lands on the banks of the Nile in order to improve the efficiency of his tax collectors. The Neo-Sumerian king Gudea of Lagash, who reigned from 2144 to 2124 BCE, is represented in a statue holding a measuring stick (now displayed in the Louvre). The milestones alongside Roman roads indicated their distance from Rome. The Greek goddess Nemesis is portrayed with a scale and a measuring rod. In the Bible, "A just balance and scales belong to the LORD; All the weights of the bag are His concern" (Prov. 16:11). Official measurements, and the ability to make them widely known and adhered to, are symbols of power, demonstrating bonds with the divine. They express a sense of belonging and mutual trust.

King Gudea held a measuring stick in his lap. Balances weighed the hearts of deceased Egyptians and determined their fate in the afterlife. The Most Serene Republic of Venice hung plaques in its markets picturing the minimum lengths required for the various species of fish to be bought and sold, to ensure the protection of young fish and therefore the environment. In modern times, prototypes of standard units of measurement—like the meter or the kilogram— were kept in capital cities close to the seats of central governments. Measurement is power but also mutual trust. It is thanks to the existence of standard prototypes maintained by trusted institutions that when we go to buy something by weight or length, we do not feel the need to take a measuring tool along with us. Even if the person behind the airport check-in counter tells us that our bag is too heavy to qualify as a carry-on . . . , well let's admit it, the thought that the scale must have been doctored immediately comes to mind.

The system of measures is a mirror of historical events. After the fall of the Roman Empire, when Europe entered the decentralized centuries of the Middle Ages, the social and political decentralization of human communities was reflected in the progressive unraveling of common systems of measurement, which became increasingly localized. Not coincidentally, subsequent unifying moments or events were always accompanied by attempts to harmonize systems of measurement over ever-larger areas. Charlemagne tried but failed to regulate measures in the Holy Roman Empire. A few centuries later, England's Magna Carta attempted to establish rules for the measurement of volume, length, and weight for commerce: "There shall be standard measures of wine, ale, and corn (the London quarter), throughout the kingdom. There shall also be a standard width of dyed cloth. . . . Weights are to be standardized similarly."

Measures are certainly not exclusive to Western cultures. In China, as recounted by Robert P. Crease in his book *World in the Balance*, measuring processes date back to before 2000 BCE. One of the first acts of Qin Shi Huang, the first emperor of unified China, was the centralization of the system of weights and measures. Crease also describes how, as early as 1400 CE, the Akan people of West Africa developed of a system of weighing based on miniature sculptures, which functioned as figurative weights and were used in the negotiation of transactions paid in gold dust.

It would not be until the seventeenth century, with Galileo and the propagation of the scientific method, and the French Revolution of the eighteenth century, that two crucial steps were taken toward the creation of a universal system of units of measurement. The modern scientific method is founded on experiments and observations and on their reproducibility. To describe these experiments, to draw new hypotheses and theories from them, or to validate or refute existing theories, a common language is needed: the language of measures. The French Revolution took place amid a universal and antiaristocratic spirit. Liberty, equality, and fraternity could not thrive in a society dominated by partisan interests that imposed biased and confusing systems of measurement. It's thought that there were thousands of different units of measurement, which, of course, favored the few who administered them over the many who were forced to use them.

The revolution demanded a universal system, the same for everybody. This need had already been recognized in prerevolutionary France and now it found fertile terrain in which to come to fruition. The Cahiers de doléances (lists of grievances) presented to the Estates-General, urgently convened in 1789 by King Louis XVI, repeatedly asked for universal systems of measurement that were controlled by the Third Estate—the bourgeoisie and the peasants, for whom processes of measurement were an integral part of their work and sustenance. It is no coincidence, therefore, that tailors demanded "that there be the same weights and the same measures throughout the realm, as well as a unitary legal system and unitary customs fees," and that blacksmiths called for "the same weight, the same measures, the same laws."

All of this led to the establishment in Paris, in the last decade of the 1700s, of a decimal metric system, composed of six units of measurement: the meter for length, the hectare for land area, the stere for volume (equal to a cubic meter of firewood), the liter for liquids, the gram for weight, and the franc for money. Of the six, only the (kilo) gram and the meter have survived as base units in our own times; they are thus children of the revolution. The unit of length was defined in the session of the National Assembly of March 30, 1791, as one ten-millionth of the distance from the equator to the North Pole, measured along the meridian that passes through Paris. Theory gives rise to practice, but old habits die hard, and almost half a century passed before the minister François Guizot promulgated a law that formally adopted the metric system in France.

The need to overcome local and national boundaries, at least with regard to systems of measurement, spread from postrevolutionary France to an international level. On May 20, 1875, in Paris, seventeen nations signed the Metre Convention, which instituted a permanent organization to act by common agreement on all questions regarding units of measurement. From then on, metrological activity intensified, thanks to the newly created International Bureau of Weights and Measures.

■ ■ ■

The week of October 10, 1960, witnessed two important debuts. The second one took place on Saturday, October 15, at Kirchenallee 57, Hamburg. There John, Paul, George, and Ringo made their first recording together at the Akustik Studio, playing the classic tune "Summertime" by George Gershwin. The first happened on Wednesday, October 12, in Paris, at the opening of the 11th General Conference on Weights and Measures. During this meeting, the international system of units of measurement (abbreviated as SI) was defined, the first truly universal system. The long and winding road of measures finally achieved a fundamental goal. Right in the middle of the Cold War, when the political boundaries between nations were becoming more rigid, the boundaries between measurement systems were eliminated. Although many people might think that the history of the twentieth century was influenced more by that week's second debut, it is actually the first that has radically changed our dialogue with the universe.

Originally, the international system was composed of six units of measure: the meter for length, the second for time, the kilogram for mass, the ampere for electric current, the kelvin for temperature, and the candela for luminous intensity. In 1971, the mole would be added as the base unit of amount of substance, fundamentally for use in chemistry. At last humanity had a coherent architecture for measurement, whose seven basic units defined a complete and universal language for measuring not only our own small world but all of nature, from the most obscure subatomic recesses to the boundaries of the universe.

Modern society, science, and technology simply could not exist without measurement. Time, length, distance, velocity, direction, weights, volumes, temperature, pressure, force, energy, luminous intensity, power: these are just some of the physical properties that are the daily objects of accurate measurements.

Measuring is an activity that permeates every aspect of our lives. We typically take it for granted, only to realize just how crucial it is when our instruments of measurement do not work correctly or are unavailable. Without the measurement of time, there would be no alarm clocks to wake us in the morning. Without the measurement of volume, we wouldn't know how much fuel is left in our gas tank. Without the measurement of position or velocity, our trains and planes would never arrive safely. Without the measurement of our bodily functions, our health would soon be at risk. Without the measurement of electricity, none of our electronic devices would work.

Science and technology have taken giant steps forward since the French revolutionaries defined the decimal metric system. Today, an amazing number of highly precise measurements allow us to verify and assess new theories. They are the tools of Nobel Prize winners—for example, the measurement of the Higgs boson or the detection of gravitational waves. They are indispensable for research on the cutting edge in all fields of science. They have allowed us to fight back against the Covid-19 pandemic, and they make modern technologies work, whether they be satellites in orbit or the smartphones in our back pockets.

These measurements are based on the international system, whose units have a physical reference in prototypical objects or phenomena, that is, in something that is accessible to everyone. We have seen, for example, that the meter was originally defined as one ten-millionth of the distance from the North Pole to the equator. For practical reasons, it was redefined in 1889 as the distance between two notches engraved on a bar of platinum–iridium deposited at the International Bureau of Weights and Measures in Sèvres. This bar was meant to serve as the standard against which to compare every other meter produced on Earth.

The second was initially defined as a fraction of the period of the Earth's rotation, the average solar day. In 1960, however, scientists realized that this definition was not sufficiently precise, given that the length of a day varies over time, so the second was redefined in terms of the revolution of the Earth around the Sun. Just a few years after that, the second was again redefined, to make it even more precise, as a multiple of the time period of the transition between the two levels of the fundamental unperturbed ground-state of the cesium-133 atom.

Sèvres is also home to the prototype kilogram, the International Prototype Kilogram (IPK), a cylinder composed of 90 percent platinum and 10 percent iridium, about four centimeters high and wide. This prototype replaced the original French definition of the kilogram, equal to the weight of a liter of distilled water at a temperature of 4°C (32.9°F).

Despite the care with which they are conserved, bars and cylinders, being pieces of metal, change over time. The standard kilogram was made in 1889, together with five other identical copies. Compared to the copies, in little more than a century the original lost 50 millionths of a gram. That might seem like a pittance, more or less the weight of a grain of salt. But if you consider the precision required by modern science, as well as that the kilogram is part of the definition of derivative units such as the units of force and energy, this variation jeopardizes the entire international system. The deterioration of artifacts, albeit philosophically in line with the ephemeral nature of their human makers, is incompatible with the universality and certainty required by modern science. There was a risk, therefore, of falling back into a new scientific dark age with no certainty of measurement.

This risk was remedied by scientists, who, on November 16, 2018, decided to redefine the units of the international system. No longer would these be based on material objects or events; instead, they would use universal physical constants, such as the speed of light in a vacuum or Planck's constant—constants that belong to fundamental physical laws and theories. The speed of light, for example, is crucial for electromagnetism and the theory of relativity, while Planck's constant is central to quantum mechanics.

This redefinition was a true Copernican revolution. In the past, these fundamental constants were determined by measurements using the units of the international system, based in turn on material prototypes. In November 2018, it was decided to reverse this procedure and to rely on fundamental physical constants, to fix their values in immutable terms, and to define the units of measure of the international system in reference to the constants themselves. Describing the international system of measurements by relying on fundamental physical constants, rather than material prototypes, amounted to an affirmation that the natural laws governing the universe are immutable. These constants can serve as the basis of a system of measurements that is much more solid than systems based on objects or events that we can see and touch. This was an earthshaking revolution for science but also for humanity—a revolution that is still not very well known and that we are about to discover.

Seven units of measure, a hymn to nature.

ONE

The Meter

112 Mercer Street

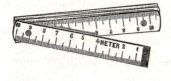The distance from 112 Mercer Street in Princeton, New Jersey, to the physics department at Lincoln University in Pennsylvania is about 148,000 meters. Put that way, it seems like a very long trek. But if we translate it into 148 kilometers, it is decidedly less imposing. Today, Google Maps informs us that, by car, the drive takes an hour and forty minutes, but in 1946 the trip was much more daunting, especially if the person traveling was in their seventies and had health problems. Considering that the reason for the trip was the conferral of an honorary degree—the kind of event that Albert Einstein did not find appealing, in view of the customary pomp and circumstance—it would not have been strange had the father of the theory of relativity declined the invitation. Lincoln, at the time, was a tiny university with an enrollment of just over 250 students.

Nevertheless, Professor Einstein accepted, and gladly so. Because—and these are his exact words, "That visit on May 3, 1946, was for a worthy cause"—Lincoln University's notoriety belied its small size. It was the first American university to confer a college degree on an African American student. Founded in 1854, it was nicknamed the Black Princeton for the ties between its founders and first faculty members and the much more famous university in New Jersey and because it was a touchstone for African American university students.

In the post–World War II years, racial segregation still plagued the Black community. Although the great majority of white Americans stubbornly refused to recognize the problem, the voice of Albert Einstein rang out loud and clear. Already in 1937, his attitude was unmistakable. He had offered hospitality to Marian Anderson, one of the most famous opera singers of the twentieth century, when she came to Princeton to sing and was denied a room in a local hotel because of the color of her skin. In 1946, in an article for the magazine *Pageant*—read primarily by a white audience—Einstein had this to say about segregation: "The more I feel an American, the more this situation pains me," adding, "I can escape the feeling of complicity in it only by speaking out." On that May 3, the Nobel Prize winner—whose emaciated face and simple ways, as a student at the time recalls, made him look almost like a figure out of the Bible—attended the ceremony at Lincoln University and gave a thank-you speech that would become famous. He spoke harshly against racism and racial segregation, calling it, "not a disease of colored people, but a disease of white people. I do not intend to be quiet about it."

It would take another nine years before the few meters that separate the front of the bus from the back of the bus—the former reserved for whites and the latter for Blacks—would mark the beginning of the modern civil rights movement, thanks to the courage of Rosa Parks, who refused to walk them. That day was December 1,

1955, and Einstein did not get to see it. He died on April 18 of that year, having played the leading role in one of the greatest revolutions of modern physics. With his theory of relativity, Einstein had reshaped not only his own discipline but the whole of human knowledge. He inspired, right up to the present, the recipients of a long series of Nobel Prizes for research originating from his theory (for which, irony of ironies, he himself was not awarded the Nobel) and became a reference point for artists, philosophers, and intellectuals, as well as a pop icon of physics.

It is only natural, therefore, that Einstein's revolution has made itself felt strongly also with regards to units of measurement. The theory of relativity describes not a specific phenomenon but rather the environment in which all physical phenomena take place: spacetime. The theory does not merely write one part of the big story that nature tells; it establishes the general rules for storytelling. Relativity is a theory about space and time, and as such, it takes precedence over all the others, which must be consistent with it.

For millennia, humanity has attempted to create for itself a global and universal system of units of measurement in order to describe and understand the world and the nature that surrounds us; a system that transcends borders and sovereignties and belongs to everyone. It is not surprising that the theory of relativity has become a milestone on the way toward the universalization of the meter, the unit with which we begin in this chapter our journey of discovery of the measures of the world.

The name *meter* itself symbolizes the principle of measurement, both for its etymology—from the Greek μέτρον (*metron*, measure)—and for having lent its name to the first international treaty on units of measure, the Metre Convention, signed in Paris by seventeen nations in 1875. This was an event that rarely appears in history books but that established a first firm foothold in the millennia-long march that started at the dawn of civilization.

From the Nile to the Tiber

Together with those for time and mass, the measure of length is one of the oldest and most familiar to humanity, since it is tied to activities fundamental for life such as agriculture. In ancient Egypt,
the measurement of terrain was an enormously important activity, so much so that the origins of geometry are attributed to Egyptian civilization and particularly to its practices of land surveying. Every year, following the season of intense rain, the Nile overflowed its banks and flooded vast amounts of land, leaving rich silt deposits. This sediment was one of the sources of the great fertility of the lands along the river. When the floodwaters withdrew, the Egyptians had to retrace the boundaries of the fields, which had been erased by the flooding.

The motive was purely mercenary, as we learn from the historian Herodotus. Referring to the decision of Pharaoh Senwosret III, who reigned around 1850 BCE, to subdivide and distribute arable land among his subjects, assigning to each a square parcel, Herodotus writes,

> They said that this king [Senwosret] distributed the land among all the Egyptians, giving each an equal square lot, and that on the basis of this subdivision, he procured revenues, having imposed a payment of an annual tribute. If the river swept away a part of the farmland, the owner, having gone to the king, reported what happened. The king then sent his functionaries to inspect and measure how much smaller the farmland had become, so that in the future the owner paid a proportion of the tribute. I believe that, as a result of this, geometry was invented and from there it passed on to Greece.

Drawing boundaries was essential for knowing whose door to knock on to collect taxes. That's why the Egyptian administration put such care into that activity and, in general, into the maintenance and continuous revision of detailed land records. The people in charge of these operations were called *harpedonapata* (rope stretchers) by the Greeks—forerunners of modern surveyors, a profession with very ancient origins! Their main working instrument was a rope. By stretching the rope between two points, they created a straight line, giving rise to the expression, still used today, "pull a straight line." They drew circles by tying one end of the rope to a fixed point, for example, a stake in the ground, and rotating the other end around it. All that attention required very precise measurements and so, naturally, there was a concerted effort to make them uniform with the introduction of units of measurement to which officials and taxpayers could refer. The most obvious choice was to use as an example something that was readily available and of "human" dimensions. What could be simpler and more universal than using some part of the human body? That is how the Egyptian cubit, corresponding to the distance of about half a meter from the point of the elbow (*cubitus* in Latin) to the fingertips, became the unit of reference not only for the subjects of the pharaoh, but also throughout much of the ancient world all the way to Rome.

In the Bible, the words cubit or cubits recur 179 times. One of its most famous appearances is in chapter 6 of the book of Genesis, where God speaks to Noah about the most celebrated embarkation of the Holy Scriptures and adds to His spiritual dictates some very practical indications: "Make for yourself an ark of gopher wood; you shall make the ark with compartments, and cover it inside and out with pitch. This is how you shall make it: the length of the ark *shall be* three hundred cubits, its width fifty cubits, and its height thirty cubits. You shall make a window for the ark, and finish it to a cubit from the top; and put the door of the ark on the side; you shall make it with lower, middle, and upper decks." Noah's ark was 150 meters long and 25 meters wide. To give you an idea of the size, keep in mind that the *Amerigo Vespucci*, the beautiful tall ship used as a training ship by the Italian navy, is 101 meters long and 15.5 meters wide.

Along the banks of the Nile, there were two versions of the cubit: the common variety, which measured about 45 centimeters, and the royal (derived from the noble arm), about 52 centimeters long, equal to the sum of the common cubit plus the breadth of the pharaoh's palm. The standard cubit was fixed in a bar of black granite, acting as a reference to which working cubits of stone or wood were produced. Examples of these have survived to the present day.

The ability to measure and the precision of the cubit played a crucial role in that gigantic feat of logistics and engineering—to use modern terminology—that was the building of the pyramids. While today the figure of 100,000 laborers cited by Herodotus is thought to be an exaggeration, more reliable estimates speak of a labor force of 10,000 people. They had neither cell phones nor computers, obviously, but they built the Great Pyramid of Giza with such precision that the exact lengths of the four sides of the base (about 230 meters each) differ by no more than 10 centimeters one from each other! That is a degree of accuracy that was not achieved 4,500 years later by the $125 million Mars Climate Orbiter, a space probe that was supposed to reach Mars and study its atmosphere and climate. When it arrived in the vicinity of the red planet, one of the monitoring computers on Earth began sending it command data expressed in English measurements. Unfortunately, the onboard computers were programmed to receive measurements expressed in the metric system and, as we know, a yard is not exactly equal to a meter . . . and that 10 percent difference was enough to make the poor Orbiter crash.

Another civilization that knew what it was doing when it came to measuring length was the Roman Empire. The road network built by the Romans was enormous. It is estimated that, at the apex of imperial expansion, Roman roads covered a distance of some 8,000 kilometers. Such a large network required a precise system of distance measurement. This need was satisfied by milestones: obelisks or posts made of stone that recorded the distance from Rome or from the nearest important city. The unit of measure for the length of the road was the mile, a name derived from *milia passuum*, which measured exactly 1,000 steps. The Roman *passus* was equal to 1.48 meters, and so the Roman mile measured 1,480 meters.

If you put the book down for a minute and try to measure your step, you might well ask yourself if the Romans had extremely long legs. Your step, in fact, typically measures 70 centimeters. The mystery is easily solved. The Roman passuum covered the distance from the point at which one foot leaves the ground to the point where the same foot hits the ground again during a march. It was not the distance from the lead foot landing to where the following foot lands, which is how we normally measure a step today.

Although the saying goes that "all roads lead to Rome," milestones did not always indicate the distance from the capital of the empire. Sometimes the stones recorded the distance from the city where the road began (*caput*, head), and sometimes the two distances appeared together. According to a study by the late classical scholar Gordon J. Laing, numeration starting from Rome is typical of the roads in central Italy, such as the Via Appia going south and the Via Emilia going north. The milestone indicating the greatest distance from Rome is found near Narbonne, France, on the Via Domitia, which went from Turin, Italy, to Spain. This marker records a distance of 917 miles from Rome and 16 miles from Narbonne. Curiously, it also shows a third distance of 898 miles, also from Rome, probably by way of a shorter route. You know how GPS navigators sometimes give you alternative routes for a given road trip? Well, they didn't invent anything new.

Allons, Enfants

Let's pause for a moment and think about the importance of the message communicated by those numbers engraved on the stones. What seem to us today to be merely practical indications useful for organizing a trip were instead a masterful message of power and inclusion, a way, perhaps as effective as sending an army, for the central government to make its presence known. Even in the most remote corners of the empire, the measure of the distance from Rome intimated not only who was in command— and who could use those roads to arrive with arms, if necessary—but also that Rome was a place where, at least ideally, anyone could go, even if they were not Romans. Rome was a center of power that was accessible to all. The milestones indicated that the government was present and that it took care of its territories. Those Roman numerals made it clear where power was located but, at the same time, gave everyone the chance to read them in the same way, no matter where they happened to be.

With the fall of the Roman Empire, the unifying force it had imposed faded away, and, not surprisingly, this had repercussions for units of measurement. For many centuries, measures of distance, among others, became more or less a local affair. Every community had its own units, which were often displayed on stone plaques in public gathering places. Many of these still survive today. In Italy, we can see them in Padua, Senigallia, Salò, and Cesena, just to give a few examples. In one of the central squares of Padua, for example, there is a stone plaque bearing the date 1277 on which the measurement units for flour, grains, bricks, and fabric are engraved. These measurements were used as standards in case of disagreements between buyers and sellers. Curiously, the location was called, in the local dialect, "Cantoon dee busie," or corner of lies: if a trader was trying to cheat somebody, the fraud would be discovered there. In France alone, there were an estimated 250,000 different units of measurement.

The new units were not all that imaginative. Many were still based on parts of the human body, typically that of the local ruler: arms, palm, feet, and so on. Naturally, the territorial extension of the validity of such measurements was much more limited than it was in ancient Egypt, and this "metric sovereignty" caused more than a few problems. Imagine what it was like for an itinerant seller of fabric or ropes. Today, we take all this so much for granted that we never think about it. The price is by the meter, and if we want a certain length of a certain fabric from a certain producer, we pay the same amount whether we buy it in one city or another or on the internet. But in those times, sellers had to recalculate the price in every village or town, and if they were less than totally honest, they had ample opportunities to cheat their customers. In general, the lack of common standards for measuring made commerce extremely difficult and often left the weaker strata of society at the mercy of the prevailing power brokers, particularly for such things as the measurement of terrain and real estate.

A cardinal element of a fully realized democracy is equal access to scientific ideas. Or at least, this should be the case. Recent events highlight the crucial nature of this access. In view of much recent experience, the conditional is obligatory. Certainly, the revolution that began with Galileo and the subsequent propagation of the scientific method were, if only indirectly, an essential component of the definition of a universal system of units of measurement. This system safeguards the interests of everyone, regardless of their status or power. Over time, the scientific community increasingly felt the need for a system that would allow for the comparison and reproducibility of the results of their experiments and observations, which, from Galileo onward, came to be recognized as the engine of scientific progress.

Several centuries were to pass, however, before the French Revolution, with its universalistic and antiaristocratic aspirations, pushed for a radical change in the system of measurement. From local—and thus not very standardized—systems, which, in commerce, often favored the few who managed them and who reaped lavish rewards from the generalized confusion, the revolutionaries wished to establish a universal system that was the same for everyone. It is not a coincidence that in the last decade of the eighteenth century, the decimal metric system, precursor of the current international system, was born in Paris.

The revolutionaries wanted to free themselves completely from the old temporal and religious powers. They attempted to introduce a decimal calendar that made it difficult to keep track of religious holidays and Sundays in particular. This experiment was not all that successful. On the other hand, two other revolutionary novelties not only survived but became fundamental to what was to become the current metric system: the kilogram and the meter. The meter was defined at the National Assembly session of March 30, 1791, as one ten-millionth of the distance from the North Pole to the equator, measured along the meridian that passes through Paris. Theory soon became practice. Two scientists, Jean-Baptiste Delambre and Pierre Méchain, were charged with measuring physically an arc of the terrestrial meridian passing through Paris. They chose the arc between Dunkirk, France, and Barcelona, Spain, equal to approximately one tenth of the distance from the North Pole to the equator. This section of the meridian has the great advantage of being for the most part flat. The two scientists set out on their expedition in 1792. Delambre measured from Dunkirk all the way to the cathedral of Rodez, France, while Méchain started from Rodez and arrived in Barcelona. They thought they would be able to do it in one year, but it took them six, in an enterprise that sometimes reached epic proportions and that was made even more difficult by its having to be carried out in a Europe turned topsy-turvy by the revolution.

In 1798, they brought their results to Paris, where they then became the basis for the definition of the length of the meter. This was given material form in a platinum bar called the *mètre des archives*. The bar was deposited in the National Archives on June 22, 1799, as the standard. A number of copies were made for practical use. To make the new unit of measure familiar to the population, samples of the meter were fixed in various places throughout Paris. Today it is still possible to see two of them, one at 36, rue de Vaugirard and one at 13, place de Vendôme.

Old habits die hard, and the introduction of the new system met more than a little resistance from the people, who continued to use the old units. Finally, in 1812, Napoleon repealed the law that imposed the metric system. In an odd parallel with the fate of the powerful, after the fall of Napoleon, the traditional measuring system was replaced in 1837 by a law that went into effect in 1840 and returned the metric system to the fore in France. However, it would take until the middle of the nineteenth century for the metric system to establish itself firmly in France and begin its conquest of the rest of Europe.

On July 28, 1861, with passage of Law 132, the decimal metric system was introduced in the Kingdom of Italy. There, too, its application was far from immediate, and the central government pressured mayors to promote the implementation of the new system by local populations. Specially designed plaques with tables of equivalence were displayed in public places. Public schools also played an interesting role in promoting metric literacy. A glance at the elementary school curriculum for 1860, for example, reveals instructions such as the following: "To these notions the teacher shall add a brief explanation of the metric system, teaching pupils the names of the new measures, explaining in detail what is meant by a meter, how all the other units of measure derive from it, and what the value is of each." And "Teachers of the fourth year and of preceding years are reminded that the most important subjects of elementary instruction are the catechism and religious history, Italian grammar and composition, arithmetic and the decimal metric system. To these subjects, therefore, they shall direct most of their attention and consecrate most of the time at their disposal in the school."

By now, the road ahead was clear, and step by step (or rather meter by meter), the revolutionary dream came true. On May 20, 1875, in Paris, seventeen nations signed the Convention du Mètre, the treaty that established a permanent organizational structure that allowed contracting countries to act in common accord on all questions related to units of measure. Furthermore, the treaty also instituted the General Conference on Weights and Measures (Conférence générale des poids et mesures, CGPM) as an international diplomatic organization responsible for the maintenance of an international system of units of measure in harmony with the progress of science and industry.

That same period saw the institution of an intergovernmental organization known as the International Bureau of Weights and Measures (Bureau international des poids et mesures, BIPM). Located just outside Paris, in Sèvres, the BIPM is the organism through which contracting states act on questions of metrological importance. The bureau also serves as both the scientific and literal custodian of the international system of units. Indeed, the international prototype meter is kept on the premises of the BIPM: a bar of platinum–iridium with a special X-shaped cross-section (the Tresca section, named for its inventor, Henri Tresca) to resist possible distortion. In 1875, the length of the meter was defined as the distance between two lines engraved in the bar (the so-called line standard) to avoid problems owing to eventual wear on the ends. Thus, the bar itself was longer than a meter.

This piece of metal would constitute the standard against which all meters around the world would be calibrated. Of course, this process required intermediaries: accurate copies of the prototype meter in Sèvres were distributed to each contracting state of the Metre Convention. The American copy is bar number 27, received by President Benjamin Harrison on January 2, 1890.

The Beginning of the End

In the eyes of those who conceived of the meter standard in 1875, the extraordinary solidity of the simple, robust bar of metal guaranteed its long life as a worldwide standard. As it turned out, however, just as the bar of platinum–iridium was being fused and put into use, physics entered a new era that would turn the world on its head and send that illustrious metal piece of the revolution's legacy into retirement. The last decades of the nineteenth century and the first decades of the twentieth were to see a sequence of discoveries that would become the basis of modern physics and technology.

Take, for example, the understanding of electromagnetism. The Scottish scientist James Clerk Maxwell could never have imagined the practical impact of what he published in his *Treatise on Electricity and Magnetism* in 1873. Consider that Maxwell's four equations simply and elegantly describe all the phenomena and the technology related to classical electromagnetism and, in particular, to electromagnetic waves. These phenomena range from the rainbow to electric cars, from the cell phone to the blue sky, from the washing machine to the elementary particles in the cyclotrons at the CERN (European Organization for Nuclear Research) in Geneva. Nor did the German physicist Heinrich Hertz, who was the first to demonstrate experimentally the existence of Maxwell's electromagnetic waves, have any idea of their future impact. It has come down to us that Hertz commented on his discovery as follows: "They will not be of any practical use. Mine is just an experiment that proves that Maestro Maxwell's theory is correct. Put simply, we have these mysterious electromagnetic waves that are there and that we cannot see with the naked eye." He was then asked, "But what is going to happen then after your experiment?" It seems that Heinrich Hertz responded modestly, "Nothing, I imagine." He certainly cannot be blamed for his lack of imagination. At the time, it really would have been impossible to predict the multitude of uses that would later be made of electromagnetic waves, for communication, for travel, for cooking, in medicine for the diagnosis and treatment of various illnesses, and in many other fields as well.

Those decades also saw enormous progress in the understanding of the structure of matter, discoveries that would pave the way for modern atomic theory: the discovery of X-rays by Wilhelm Röntgen in 1895 and of the photoelectric effect in 1887 by Hertz (an effect described completely by Einstein in 1905 and that earned him the Nobel Prize in 1921). Ten years later, in 1897, the electron was discovered by J. J. Thomson. These were fundamental milestones on the road to understanding how the matter that surrounds us is made of microscopic particles known as atoms, which, in turn, are constituted by a nucleus made of protons and neutrons and by electrons placed outside that same nucleus. A few years later, this road opened up the quantum revolution.

The precision guaranteed by two lines engraved on a bar of metal, albeit a noble metal, could hardly have remained sufficient to measure the new world that physics was in the process of discovering. Accurate as it was, that object was quickly becoming a vestige of a bygone age, incapable of holding up under the ever more pressing demands of the new physics, which was abandoning the human dimension to venture into the unimaginably small and the infinitely large. In a few decades, the range of the size of the objects studied by physics expanded, literally beyond measure, from the fractions of a billionth of a meter of Niels Bohr's atom to the hundreds of thousands of billions of billions of kilometers of the distance from Earth to the galaxies on the edges of the universe studied by the astronomer Edwin Hubble.

Paradoxically, the fate of the international prototype meter was already sealed when it came into being. The artifact in Sèvres became the victim, on the one hand, of the continuing demand for greater precision in measurement spurred by new discoveries in physics and technology and, on the other, by the globalization of both. Born in a nineteenth-century Eurocentric context, where the distance between centers of knowledge were comparatively short, in the twentieth century the meter collided with a scientific world on which the Sun never set. The copies of a material object such as the prototype meter, however accurate, were nevertheless perishable, and they could not be everywhere they were needed at the same time. Beyond that, they became increasingly insufficient at measuring the new and ever more extensive world that was being discovered.

The prototype meter could still work well, for example, in measuring the dimensions of a soccer field on which one of the stars of the early 1900s was Harald Bohr, a member of the Akademisk Boldklub of Copenhagen and of the Danish national team that won the silver medal at the London Olympic Games in 1908. But it was certainly not enough for his more famous brother Niels—perhaps the only case of a family with the scientist who became more famous than the professional soccer player—who in 1913 published his article "On the Constitution of Atoms and Molecules" in the *Philosophical Magazine*. In this study, Niels laid the basis for the modern quantum theory of the atom, or quantum mechanics. He described the atom as a microscopic solar system the size of roughly one ten-billionth of a meter, with the nucleus at the center and the electrons orbiting around it. He also foreshadowed, with an ingenious intuition, the quantization of energy.

Among the four postulates that form Bohr's theory of the atom, one details how an atom emits electromagnetic radiation with a well-defined level of energy. This occurs when an electron, initially moving along a given orbit, changes its motion discontinuously and jumps into a different orbit with lower energy. According to Bohr, the frequency of the emitted radiation is equal to the difference between the two energy levels divided by Planck's constant, another fundamental constant of physics that we will encounter later on. Because of the transitions of its electrons between orbits, every atom can emit electromagnetic radiation with a limited and preestablished set of energy values or, if you will, colors. This set of energies is the atom's spectrum, and each component of the periodic table has its own, a sort of palette for each atom. To use an example that will be understood by everybody, when you are cooking pasta and some drops of boiling water spill out of the pot and drip down onto the gas flame, you see that the flame turns yellow, a color that belongs to the emission spectrum of sodium. However, the experiment will not work if you have forgotten to add salt to the water: the sodium comes from the salt.

It was, in fact, an atom that sent the prototype meter into retirement. Less than a century after its introduction, it was abandoned in favor of a definition that was decidedly less tangible.

In 1960, the meter was redefined thanks to Superman. Well, not exactly, but part of the blame belongs to krypton, an element of the periodic table with the atomic number 36, which inspired the creators of Superman when naming the planet of the superhero's origins. Krypton is an inert gas, frequently used to emit light in so-called neon lights (which are not always made of neon . . .). Thanks to progress in optics, by this time the length of visible radiation waves emitted by atoms was measured with much greater precision than that with which the distance between the lines on the metric bar was determined (lines whose widths, in any case, may be quite small but are not negligible). It was decided, therefore, to define the meter as equal to 1,650,763.73 wavelengths of the radiation emitted during a particular energy transition in the krypton 86 atom. The meter takes on a reddish orange color, exactly how that particular radiation appears to the eye.

The change, established by the General Conference on Weights and Measures on October 14, 1960, is of epochal significance because the definition of a meter passed from a human-made artifact, such as a metal bar, to a natural phenomenon, the light emitted by an atom. The ephemeral nature of human creations was replaced by the eternity of nature. The use of krypton initiated a reliance on nature over human artifacts, leading to the recent revolutionary redefinition of the fundamental units of measure based on universal physical constants.

Krypton's notoriety, however, was destined to remain tied more to Superman than to the meter. Indeed, only about twenty years later, in 1983, a new star of metrology took the stage, and it was destined to stay there for a long time: the speed of light.

A New Relativity

In our collective imagination, physics is often represented by two stereotypes. The first is of wacky scientists

$$C = \text{const}$$

who live with their heads in the clouds—preferably with some characteristic quirk like rumpled fuzzy hair or weird clothes and multicolored socks on display on every occasion—and have blackboards chock-full of incomprehensible-looking equations. The episode of Albert Einstein's visit to Lincoln University, however, tells us that, much more often, physicists are women and men like everybody else, who live, for better or worse, in the real world and sometimes try to influence it. In 1943, another pioneer of nuclear physics, the German chemist Fritz Strassmann (who, together with Lise Meitner and Otto Hahn, discovered fission) hid the Jewish musician Andrea Wolffenstein for several months in his home in Berlin to save her from deportation. Strassmann was a proud opponent of Nazism. After resigning from the Society of German Chemists, which had come under the control of the Nazi Party, he declared, "Despite my affinity for chemistry, I value my personal freedom so highly that to preserve it I would break stones for a living," which made it rather difficult for him to find work. For what he did for Andrea Wolffenstein, Strassmann is remembered today in Yad Vashem's list of the Righteous Among the Nations.

The second stereotype is the incomprehensible-looking equations. To be sure, physics is not an easy discipline, but sometimes many of its most revolutionary findings are represented by equations that are simple and elegant, such as, for example, the following:

$$c = \text{const.} \tag{1}$$

It may be hard to believe that this expression encapsulates a big piece of Einstein's special theory of relativity, but it's true. Let's take a look at it.

We'll start with the star player of this equation: light. The term *light* must be handled with care, because it could appear to be reductive. In our everyday human experience, in fact, we associate light with vision, but for physicists, the noun has a much broader significance. In essence, the light that we see is an electromagnetic wave that travels through space. Like any other wave—the waves of the ocean, sound waves, or the wave of fans at the football stadium—electromagnetic waves transmit information by way of a periodic modification of a certain physical dimension. In the case of sound waves, this is air pressure; for the waves of the ocean it's the position of the water; for the stadium wave, it's the position of the fans in the stands. For electromagnetic waves, what is modified is the electromagnetic field, an invisible but very real entity, which physicists use to describe some properties of space and matter.

These electric and magnetic properties were also known to the ancients. The ancient Greeks knew, for example, that a piece of amber—in Greek, ἤλεκτρον (*electron*)—rubbed with silk or fur would attract threads or pieces of straw, and that a stone present in nature (magnetite) attracted iron. Ancestors of the compass, which even back then exploited the interaction of a material with the terrestrial magnetic field, were known in China during the Han dynasty (second century BCE to second century CE).

It was not until the nineteenth century, however, that the fundamental laws of electromagnetism were fully understood and endowed with a theoretical apparatus. The key was the introduction of the electromagnetic field, made of an electric component and a magnetic component. As we have seen, Maxwell expressed in four fundamental equations the relationships between electric and magnetic fields and their sources, respectively, electric charges and currents. During the last years of the nineteenth century and the first years of the twentieth, in parallel with developments in theory and research, advancements were made in practical uses of electromagnetism. Cities began to be illuminated by electric light, distances were shortened with the first telegraphic and radio transmissions, and electric engines began to power factories.

Albert Einstein began his research in this innovative and dynamic context, which, for a physicist, was also problematic. Like all physicists, Einstein had been trained in the classical mechanics of Galileo and Newton. For more than 250 years, this body of knowledge not only worked but yielded admirable results such as those, for example, related to the motion of celestial bodies. The theater of Newtonian mechanics was three-dimensional space, described by a system of coordinates: three perpendicular axes with a common origin that made it possible to identify any point with three numbers. An example of such a coordinate system—in this case, on a plane—is the game board for Battleship, where each position is identified by a letter and a number. Each of us can choose the frame of reference, and its origin, which we like best, something that we normally do when we measure distances typically from the place we are in. I live in Venice, and when I talk about Padua, I generally say it is 38 kilometers from here. More rarely, I might say it is 984 kilometers from Giessen, Germany, and 43 from Rovigo, Italy. On the stage of classical physics, phenomena unfold in time, which flows steadily and unchangeably, always the same for everybody and with a clear and universal demarcation between past and future. Space and time are rigidly separated.

The cornerstone of classical mechanics is the principle of Galilean invariance, or Galilean relativity, which holds that the laws of physics always have the same form in frames of reference that are moving at a constant velocity with respect to one another. In other words, if we were playing pool in the living room of our home or on a train traveling at 300 kilometers per hour, the laws that describe the motion of the balls remain the same, and when we observed their motion, we wouldn't be able to tell if we were moving or standing still. The only caveat is to be careful to transform the coordinates of each point when we move from one frame of reference to another, and for this purpose, Galileo provides us with precise formulas. With physics experiments, therefore, it is not possible to determine whether a means of transportation is stopped or moving at constant velocity. In his *Dialogue Concerning the Two Chief World Systems*, Galileo states this very clearly by describing an ideal experiment to be carried out on a ship below deck, from where it is impossible to see outside (and, therefore, to tell if the ship is moving by observing, for example, the coast): "Have the ship proceed with any speed you like, so long as the motion is uniform and not fluctuating this way and that. You will discover not the least change in all the effects named, nor could you tell from any of them whether the ship was moving or standing still."

The classical mechanics of Galileo and Newton was very elegant and coherent. But now there was a new science, electromagnetism.

It, too, like Maxwell's equations, was quite elegant and had multiple practical applications. The problem was that electromagnetism and Galilean transformations were not compatible. In passing between two inertial frames of reference, with one in relative motion at constant velocity with respect to the other, some of the fundamental laws concerning electric and magnetic fields came to be modified. This seemed very disturbing. Einstein himself admits in his *Relativity: The Special and the General Theory* that "the question of the validity of the principle of relativity became ripe for discussion, and it did not appear impossible that the answer to this question might be in the negative."

His solution to the problem was the special theory of relativity. Einstein started from the principle of relativity, which he not only assumed to be true but even extended from mechanics alone to all of physics, including electromagnetism. But he added a second postulate to the general theory—summarized in the few letters of equation (1) above: that in a vacuum light always travels at the same velocity c in all inertial frames of reference. Not such a big deal, it would seem. Yet it revolutionized physics.

Let's look for some help in an example. Suppose we are on the deck of a passenger ship that is moving at a speed of 20 kilometers per hour, and we go for a run, let's say at 10 kilometers per hour, in the same direction as the movement of the ship, toward the prow. Our speed of 10 kilometers per hour is measured with respect to a frame of reference that is moving with the ship. To a friend standing on the shore, we are running at 30 kilometers per hour, since the velocity produced by our legs is added to the velocity of the ship. Well, for light, it doesn't work that way. When we measure the speed of light, we always come up with the figure of 299,792.458 kilometers per second, regardless of the frame of reference against which it is measured.

Accepting that the speed of light is a universal constant that does not depend on the frame of reference has, as an immediate consequence, led to a profound redefinition of the concepts of space and time. In Galilean relativity, lengths stayed the same while passing from one frame of reference to another. The length and width of the pool table stayed the same, whether they were measured on the train or in the living room, just like the length of the ship and the line along which we went for our run. Moreover, for Galileo time was absolute; it flowed the same way everywhere.

With Einstein, everything changed. To make the principle of invariance consistent with the universality and constancy of the speed of light in a vacuum, Einstein modified Galilean transformations. First of all, for length. In the new special relativity, lengths in the direction of motion become shorter in the frames in motion. Next, for time. In Galilean transformations, time is an independent parameter with respect to the frame of reference and with respect to position. According to Einstein, time loses its impartial status: time gets mixed up with space; it becomes relative. Time passes more slowly in systems in motion; it dilates.

We are normally unaware of all of this because the shortcomings of Galilean transformations become significant only when the relative velocity of the frames of reference approaches the speed of light. In normal human experience, approaching the speed of light is not something that happens very often, which is why Galilean relativity works fine most of the time, but it is incomplete.

The speed of light, or more technically the speed of electromagnetic waves in a vacuum, becomes one of the pillars of our scientific knowledge of the world. An unchangeable feature of nature. A universal constant. From that moment on, the theory of relativity, which is based on that constant, becomes a meter for all of physics, an indispensable condition. As Einstein wrote in *Relativity*: "This [the space-time transformation] is a definite mathematical condition that the theory of relativity demands of a natural law, and in virtue of this, the theory becomes a valuable heuristic aid in the search for general laws of nature. If a general law of nature were to be found which did not satisfy this condition, then at least one of the two fundamental assumptions of the theory would have been disproved."

From the Earth to the Moon

After his visit to Lincoln University, Einstein lived only nine more years. He died in 1955, and fate denied him the chance to see the invention of one of the premier scientific instruments for the study of light: the laser. The first prototype was made in 1960. The laser produces a well-collimated and monochromatic beam of light,  which is to say that the light emitted is of a precise color. In more technical terms, all of the electromagnetic energy that composes the beam has the same well-defined frequency and therefore the same energy, and this allows it to be monitored with accuracy. Even when it makes a round-trip voyage to the Moon.

The laser's monochromatic property and the fact that the speed of light is a universal constant are the basis of one of the laser's first spectacular applications: establishing with precision the distance between the Earth and its satellite. This enterprise was accomplished in 1962 by an Italian physicist, Giorgio Fiocco, who at the time was working at MIT. In essence, Fiocco shot a laser beam toward the Moon and measured the light that came back after being reflected off its surface. With great experimental tenacity, Fiocco and his colleague, Louis Smullin, searched for the packets of light that were reflected directly off of the lunar surface. Given the extremely weak intensity of the reflected light, this was not an easy task; but between May 9 and May 11, 1962, they succeeded, and their results were published on June 30, 1962, in *Nature*. They paved the way to a multitude of other applications. Since the speed of light is well known and constant, by measuring how much time light takes to return to Earth after being reflected off the Moon—for the record, about 2.5 seconds—you obtain a remarkably accurate measurement of the distance between the Earth and the Moon, on average a distance of 384,400 kilometers.

Fiocco's measurement is also conducted in our day to measure the Moon's distance from Earth, but with the assistance of instruments left there by Apollo astronauts. This was an experiment entitled Lunar Laser Ranging. Because of the relationship between its relative simplicity and the huge amount of information it has produced, it has been called the most successful experiment of the Apollo 11 mission. The component transported to and left on the Moon is, in fact, a mirror: a square panel with sides measuring about 50 centimeters oriented toward Earth. Fixed to this panel are a hundred or so retroreflectors, special reflectors able to reflect light with great efficiency back in the same direction it comes from (the same principle applied by reflectors on bicycle wheels). The light in question is the light "shot" from Earth with a laser, and the mirror functions as the target.

Shooting from the Earth and hitting an object not much bigger than a pizza box placed on the Moon may seem like something out of science fiction. Yet the scientists from the Lick Observatory in California could already do that, with the aid of a powerful telescope, just a few days after the landing on the Moon. This was no mean feat, given the light's voyage of more than 700,000 kilometers and especially considering that the light beam reaches the Moon with a diameter of about four kilometers and that only the light that hits the little mirror is useful for the measurement. No wonder scientists have written that aiming at a mirror on the Moon is like using a rifle to hit a coin at a distance of three kilometers! With Lunar Laser Ranging, the distance from Earth to Moon was determined with a precision on the order of one centimeter, a margin of error of less than one ten-billionth. The quality of the light emitted by the laser makes it possible to reach levels of precision that were previously unimaginable, even—as we have just seen—for the measurement of length. So it was the laser that brought about, in 1983, the most recent modification of the definition of the meter, the last probably for a long time. It was first among the new discoveries that led, in 2018, to the complete redefinition of the international system of measurement based on universal physical constants.

"*c*" as in Universal

"*c*"—a simple letter that holds within itself a universal property of nature. A property of everyone and for everyone, intangible and immutable, totally exempt from the inevitable decay of human experience. Is it any wonder that it was precisely this property that was chosen for a universal definition of the symbolic unit of the decimal metric system?

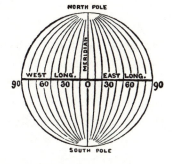

Before we arrive at the last act of the millenary history of the unit of length, it is worth making a small digression to recount the origins of *c*. Indeed, it is quite legitimate to ask why "*c*" and not "*a*," "*b*," or something else. Both Maxwell and Einstein, in his first article in 1905, had used "*V*," for example. Other physicists used "*c*," however, and it was this definition that took hold, so that even Einstein himself converted in 1907. In reality, there is no clear answer to this curious question. One school of thought ascribes the origin of the choice of "*c*" to *constant*, based on the universality of its value. Another explanation ties the choice to the Latin term *celeritas* (velocity). As of today, despite numerous studies, the ambiguity is still with us. All things considered, a little aura of mystery about a celebrity like *c* isn't such a bad thing.

The long history of the meter concludes, at least for now, in 1983, when the last act was staged. At the 17th General Conference on Weights and Measures, it was determined that "the present definition does not allow a sufficiently precise realization of the meter for all requirements" and "that progress made in the stabilization of lasers allows radiations to be obtained that are more reproducible and easier to use than the standard radiation emitted by a krypton 86 lamp" (which was the basis of the redefinition in 1960). Above all, however, it was noted that "progress made in the measurement of the frequency and wavelength of these [laser-produced electromagnetic] radiations has resulted in concordant determinations of the speed of light whose accuracy is limited principally by the realization of the present definition of the meter" and that "there is an advantage, notably for astronomy and geodesy, in maintaining unchanged the value of the speed of light recommended in 1975 by the 15th CGPM in its Resolution 2 (*c* = 299,792,458 meters per second)."

In other words, this means that between the meter of French revolutionary days and the speed of light, science holds the second to be more reliable. Another great tribute to the genius of Einstein. The exigencies of modern science and technology require levels of precision in the measurement of length that even the meter defined in terms of the radiation emitted by krypton can no longer ensure. Rather than chasing after new definitions that might allow ever more precise determinations of *c*, the 1983 convention preferred to establish a fixed point. The speed of light is defined, once and for all, at the precise value known at the time, 299,792,458 meters per second, and the meter is defined indirectly in relation to *c* and the definition of the second.

Since velocity corresponds to the ratio between the measure of the length and that of the time employed to cover it, the length of a meter is defined, thus and simply, as the distance traveled by light in a fraction of a second equal to 1/299,792,458. The definition of the meter is, therefore, indirect, and is based on the definition of the second, which, as we shall see, is measured, thanks to atomic clocks, with much greater precision than that with which the meter could be measured directly.

The era of definitions based on human artifacts came to an end for the meter in 1983 and has begun to end for all the other units of measure. With the meter, humanity has given its initial approval to a system that does not depend on physical objects but is based entirely on the speed of light and other universal constants of physics. These constants are fundamental for a series of well-consolidated scientific principles, and they represent the backbone of our continuously growing knowledge of the laws of nature.

A system of measurement that could be, finally and truly, for all time and for all people.

TWO

The Second

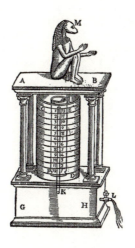

A Moment of Madness

There are those who do it because a moment of madness can happen to anyone. For the more fortunate among us, such a moment occurs while they're sleeping, but to some it happens when they're awake, inside a store. For others, it's a matter of a thirst for revenge, to strike back for a suffered wrong. Or mere masochism. Or even a flash of lucid cruelty. Whatever the motivation, there are lots of us who purchase, and who then, perhaps, give as a present to friends or enemies, one of the world's most feared ornamental trinkets: a snow globe, the mythical glass ball that, when you shake it, gives you the impression that snow is falling inside it. The operating principle is always the same: the sphere is full of a transparent liquid surrounding a small three-dimensional artifact that reproduces a landscape. The scenes are often related to Christmas, but not always. Some represent famous monuments, puppets, cartoon characters, or religious scenes. All that beauty is the fruit of the ingenuity of Erwin Perzy, a Viennese manufacturer of surgical instruments, who made the first snow globe in 1900. Inside was a reproduction of the Mariazell Basilica, with snowflakes made from microscopic fragments of grated rice. Today there is even a museum in Vienna in commemoration of the inventor that houses a collection of his most prized pieces.

A smidgen of hypocrisy prevents many of us from admitting that we have at least been tempted to buy a snow globe. It prompts others to embellish the purchase with a cultural flourish, citing the scene from the Orson Welles film *Citizen Kane* dedicated to our little knickknack. Nevertheless, the success of this souvenir is confirmed by the data. According to a study conducted a few years ago and reported in numerous newspapers, the snow globe is the most frequently confiscated object at security checkpoints in the London City Airport. Many of them, in fact, contain a quantity of liquid in excess of the amount permitted for carry-on luggage, and so they end their journeys ingloriously in the hands of security personnel, saving from harm, in all likelihood, an old friendship or a nascent romance. Other, much more banal, items fill out the roster of items seized at the London airport: cosmetics, bottles of alcoholic beverages, tennis rackets, handcuffs (!). Not even one atomic clock, it seems.

In 1971, by contrast, Joseph Hafele and Richard Keating, traveling in an era when airport controls were far more relaxed, had no problem carrying an atomic clock onto their commercial airline flight. Pictures from the time show the clock as a rather cumbersome parallelepiped, about the size of an average refrigerator, that managed to make more than one around-the-world voyage in the company of its two fellow travelers.

Hafele and Keating were a physicist and an astronomer, respectively, and the airplane voyage of the atomic clock was a crucial experiment for verifying with macroscopic clocks the modification of time predicted by Einstein's theory of relativity, whether caused by movement—in the special relativity theory—or by gravity, in the general theory. As reported in the abstract of the famous article they published immediately after their voyage in the prestigious journal *Science*, the experiment was a success:

> Four cesium beam clocks flown around the world on commercial jet flights during October 1971, once eastward and once westward, recorded directionally dependent time differences which are in good agreement with predictions of conventional relativity theory. Relative to the atomic time scale of the U.S. Naval Observatory, the flying clocks lost 59 ± 10 nanoseconds [billionths of a second] during the eastward trip and gained 273 ± 7 nanoseconds during the westward trip, where the errors are the corresponding standard deviations. These results provide an unambiguous empirical resolution of the famous clock "paradox" with macroscopic clocks.

On an airplane traveling at 900 kilometers per hour, a day is longer by a few dozen nanoseconds. That seems like a negligible amount, but even a simple smartphone makes dozens of arithmetical operations in that time. . . .

Easier to Get the Philosophers to Agree?

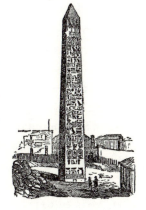

No other physical quantity has had more exposure beyond the confines of science than time. That is not surprising, given that the passing of time is associated with the most precious thing we have, our lives. Time consoles us, torments us, gives us hope, teaches us. We live in the present moment squeezed between past experience and future expectations. Time permeates our existence. We try to come to terms with it but without much hope of succeeding. Time is so evidently wound up with our lived experience that we can scarcely imagine how hard it is to describe it. Augustine of Hippo came to understand this as early as the fourth century CE, when, in response to the question "What is time?" he declared, "If nobody asks me, I know; if I try to explain it to someone who asks me, I don't know." The Nobel Prize winner Richard Feynman tried to give an answer to this dilemma when he wrote in his lectures, "What really counts is not how we define it, but how we measure it."

The pragmatic approach of the American physicist reflects what humanity has done with time since the dawn of civilization. Unable to control it, and probably long before asking themselves what it was, people have at least tried to measure it. At the start, by exploiting natural cyclical phenomena: the alternation of day and night, the seasons, or the phases of the Moon. Common to all techniques of measuring time is the use of some periodic phenomenon, something that repeats regularly. For example, the alternation of light and dark in the course of a day or, as we shall see, the succession of minimums and maximums in the emission of radiation by an atom.

Calendars were introduced as far back as the Neolithic period. Some experts even claim that the incisions on a thin plate of mammoth ivory, found in France and dating back to 30,000 years ago, are recordings of all the lunar phases in a one-year period. To be sure, many objects throughout the world are contenders for the title of the first calendar. One of the first pocket calendars—probably used in agriculture—is about 10,000 years old and was found in the Alban Hills outside of Rome. Twenty-eight notches etched on a small rock are thought to denote the number of days in a lunar month.

Another common feature of all the techniques for measuring time is the projection of time onto something visible. The sensation of the passing of time, if only because of aging, is part of our common mentality. In order to measure the time that passes, however, we need to make it tangible: the movement of a shadow or a needle, the sound of church bells, a quantity of sand in an hourglass, the shortening length of a candle or a burning stick of incense, the smell of fresh-baked bread in front of a bakery.

As with measuring length, ancient Egypt was also in the avant-garde when it came to measuring time, along with Sumer-Babylonia. To the Sumerians, we owe the invention of the sexagesimal numeral system, with which, today, we measure the seconds in a minute and the minutes in an hour. The moving shadows cast by obelisks could well have measured the passing of time during the day as well as the change of seasons. The famous Luxor Obelisk, which today adorns the place de la Concorde in Paris, is more than 3,000 years old. Curiously, it was given to the French in 1830 by Muhammad Ali, the Ottoman governor of Egypt and the Sudan, in exchange for a mechanical clock. That was probably not such a great deal.

Egyptian ingenuity went even further, with the development of water clocks: hourglasses or containers from which, through a calibrated opening, a liquid flowed into a graduated vessel. The quantity of water that flowed out was proportionate to the elapsed time.

While today Switzerland is the world's exporter of fine timepieces, in ancient times Egypt was a major player in timekeeping devices. A silent witness to Italian political maneuvers, the Egyptian obelisk in Rome's Piazza Montecitorio, home of the Italian parliament, was originally the Horologium Augusti, a grandiose sundial commissioned for the area near the Ara Pacis in the Campus Martius by Emperor Augustus and completed in 9 BCE.

Solar and water clocks were widely used in ancient Rome, but the very development of Roman civilization—and with it the growing need to measure time—highlighted the imprecision of the instruments then available. The inaccuracy of sundials was so notorious that the philosopher, statesman, and satirist Lucius Annaeus Seneca declared (philosophers and their friends will please pardon us), "Horam non possum certam tibi dicere (facilius inter philosophos quam inter horologia conveniet)" (I cannot tell you the exact hour [it is easier to reach agreement among philosophers than among clocks]). Aulus Gellius, in his *Noctes Atticae* (Attic Nights), puts into the mouth of playwright and humorist Plautus what would today be classified as outright antiscientific invective (here in the Loeb Classical Library translation by John C. Rolfe):

> The gods confound the man who first found out
> How to distinguish hours! Confound him, too,
> Who in this place set up a sun-dial
> To cut and hack my days so wretchedly
> Into small portions! When I was a boy,
> My belly was my only sun-dial, one more sure,
> Truer, and more exact than any of them.
> This dial told me when 'twas proper time
> To go to dinner, when I had aught to eat;
> But nowadays, why even when I have,
> I can't fall to unless the sun gives leave.
> The town's so full of these confounded dials
> The greatest part of the inhabitants,
> Shrunk up with hunger, crawl along the streets.

The demand for accuracy did not find many answers. On the contrary, with the fall of the Roman Empire, the evolution of the measurement of time in Europe remained substantially blocked until the Middle Ages, when the formation of new communities rekindled the need to measure time. Mechanical clocks, installed in the towers of public buildings, became instruments of community identity. Among the most famous was the Clock Tower overlooking Saint Mark's Square in Venice, commissioned in 1493. A true revolution in the measurement of time, however, would have to wait for the dawn of modern physics.

Time Swings Like a Pendulum Do

One of them jumps half-naked out of the bathtub screaming, "Eureka!" Another gets hit on the head by an apple. Still another gets hypnotized by the rocking of a ceiling lamp in the cathedral of Pisa. And the last, chronologically, but certainly not the least, sticks out his tongue at photographers. Anyone could get the idea that physics is a discipline dominated by some pretty strange people and that their discoveries must be the fruit of a sort of instantaneous epiphany. Unfor-

tunately, this view is more widespread than you might think, and it is unfair to a science whose highest moments, and even its ordinary ones, are the fruit of a happy combination of study, discipline, doubt, and, of course, pure improvisation. This last, however, springs—as it does in jazz—from a solid base of technical skill. In this sense, by the way, science at its best is democratic, and the scientific method is ideally a realm of "one person, one vote," provided each scientist has carried the burden of hard work and study. Training and research are the fundamental ingredients of science, as flour is for bread. There are no shortcuts.

That said, the fact remains that anecdotes are a powerful narrative tool that can be put to good use to explain a crucial discovery for the modern measurement of time: the isochronism of short swings.

We are in the last years of the sixteenth century. Galileo Galilei is a professor in Pisa and sets down the basis of the scientific method. "The method that we will follow will be to make what we say depend on what has been said, *without ever supposing* as true what must be explained." As the story goes, Galileo went to visit the cathedral in Pisa and was entranced by the rhythmic swinging of a ceiling lamp. Observing it carefully and measuring the duration of the swings by counting his own heartbeats—once again, a measurement of time by way of a repetitive phenomenon—Galileo realized that the swings all had the same duration, regardless of their amplitude, as long as they were relatively small. This is a phenomenon known as the isochronism of the pendulum. Legend has it that the lamp that was the object of Galileo's observations is the same one that can be seen today in the Pisa Cathedral. In all likelihood, that is not true, given that Galileo's observations date to a time before the construction of the lamp—the work of Vincenzo Possanti in 1587—which hangs in the cathedral's central nave.

Legends aside, the isochronism of short swings is a fundamental property for the measurement of time. Galileo intuited that by recording the number of the swings—exactly what is done by a pendulum clock—it was possible to obtain a much more accurate measurement of time than by previously available methods. Toward the end of his life, he designed the first pendulum clock. His son Vincenzo built a prototype, but the Dutch scientist Christiaan Huygens is remembered as the inventor of the pendulum clock because he produced the first functional prototype in 1656.

The application of the swinging phenomenon was extended into many areas. Toward the end of the sixteenth century, the ancestor of the metronome made its first appearance: the *chronomètre* of Loulié, named for the French musician who designed it. Up to that time, musicians used the beating of their pulse to keep time, but, obviously, the results varied considerably. Instead, the chronomètre of Loulié was based precisely on the isochronism of small swings. It was itself a pendulum with regulated distances, and thus regulated time periods, which musicians could observe. Étienne Loulié wrote: "The *chronomètre* is an instrument by means of which composers will henceforth be able to mark the true tempo of their composition; and their airs, marked according to this instrument, will be able to be performed, in their absence, as if they themselves were beating time."

From the chronomètre of Loulié, technology led to the prototype of the instrument known today as the metronome, which gives the rhythm for musical executions by way of varying beats per minute. The German inventor Johann Nepomuk Maelzel was the first to patent it, in 1815, but there followed a bitter dispute with the

Dutch inventor Dietrich Nikolaus Winkel, who claimed to have created the first design.

Use of the metronome began to spread among musicians. Before then, the tempo, or the speed with which a piece of music was played, was not specified quantitatively in the score. Rather, the tempo was indicated by qualitative adjectives such as *adagio, andante,* or *allegro vivace,* and each execution was subject to a certain degree of arbitrariness as well as the experience of the musician, albeit within the range of the musical context in which the piece was written.

The advent of the metronome made it easier to make the definition of the tempo more objective. One of its first users was Ludwig van Beethoven, who embraced it with enthusiasm, so much so that he attributed to the metronome the success of his Ninth Symphony. In the Ninth, Beethoven noted the tempo by using one of the first metronomes, which unfortunately went missing in 1921 during an exhibition dedicated to the composer in Vienna. After composing the Ninth, Beethoven went back to write in the tempo on the scores of his other eight symphonies and other previously composed works. His annotations, however, gave rise to a still raging dispute among musicians, many of whom claim that the tempi indicated by Beethoven are too fast or even dissonant. One famous example is the sonata op. 106, "Hammerklavier," which begins with a nearly unplayable notation of 138 beats per minute. Musicologists and musicians have been arguing about this for a long time and are now divided into two factions, one of which has decided to ignore Beethoven's later notations, while the other prefers to adhere strictly to the tempi proposed by the composer. Adherents of the former group have particularly raised doubts about both the authenticity and the objective validity of the annotations, suggesting possible errors in transcription or a malfunction of Beethoven's metronome.

An interesting and singular contribution to the dispute has been made by the physicist Almudena Martín Castro and the big-data expert Iñaki Úcar of the University Carlos III in Madrid. In an article published in the journal *PLOS One* in 2020, they used data science technologies to analyze thirty-six different executions of the nine symphonies, establishing that the tempi chosen by the various musicians were almost always slower than Beethoven's notations. Thanks to a mathematical model of the metronome used by Beethoven, the two researchers concluded that Beethoven or his assistant must have misread the instrument, perhaps owing to a scarce familiarity of the workings of a still totally new technology whose design had not yet been optimized and which was still lacking a cultural context that could support its use. Regardless of these mathematical conclusions, it should be noted that many gifted orchestra directors interpret Beethoven in accordance with the tempi indicated by the composer.

Music and Atoms

"Without music to decorate it, time is just a bunch of boring production deadlines or dates by which bills must be paid."

This pronouncement was made, as you have probably guessed, by another rather famous musician, Jimi Hendrix. It is an affirmation that many would agree with, though it must be said that the development of the modern system of measuring time in the fourteenth century was, in part, rooted in the need to better organize community life and economic activity. Mechanical clocks in

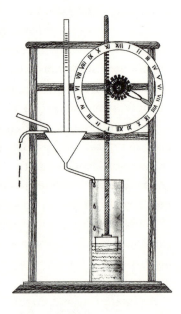

the towers of cities and towns started to spread throughout Italy in the fourteenth century, but it was thanks to the observations of Galileo and the successive practical applications by Huygens that the pendulum brought about an enormous leap forward in the accuracy of time measurement. From the earliest mechanical clocks, which were generally accurate to within about 15 minutes, the pendulum clocks of the late seventeenth century measured time to an accuracy of about 15 seconds. A further leap forward came about thanks to John Harrison, a carpenter with a passion for clock making. Between 1750 and 1760, Harrison developed a clock of reduced dimensions that was accurate to within three seconds per day and that became, for ships, a crucial instrument for the measurement of longitude, at the time a serious problem for navigation.

The technology continued to improve through the early twentieth century. In 1921, an English railway engineer by the name of William Hamilton Shortt developed an electromechanical pendulum clock, accurate to about one second per year, which became a standard for the measurement of time.

As often happens in the history of science, just when technological innovation is at its peak, the first sprouts of the next advance begin to germinate. This also occurred in the case of clocks. As Shortt's clock was becoming the standard for time measurement, Warren Marrison and J. W. Horton built the first quartz clock at Bell Laboratories. It, too, like pendulum clocks, is based on a regularly repeating phenomenon whose duration can therefore be used as a unit of measurement for time.

Quartz clocks exploit the fact that quartz itself is a piezoelectric material that can act like a tuning fork. This means that when it is shot through with electric current, the quartz crystal deforms elastically with a very regular rhythm. As the crystal deforms, it creates a weak electric signal. In a common quartz wristwatch, which sells for $30 or $40, the crystal inside vibrates 32,768 times a second, ensuring that it is accurate to within about 15 seconds per month.

As early as the 1940s, the quartz clocks used for the official measurement of time were much more accurate than a quartz wristwatch. Sufficiently insulated, they were much more robust than a pendulum clock and were not influenced by gravity, noise, or mechanical effects such as external vibrations. They were accurate to within about three seconds per year, just a little less than a Shortt clock, but their reliability and lower maintenance helped them win out as the standard for time measurement. The quartz clock remained the standard for time intervals in the United States until the 1960s.

In 1960, the second was still understood in the same way it had been for centuries. It was a fraction of an Earth day, that is, the time our planet takes to complete a full rotation on its axis. To be precise, a second was equal to one eighty-six thousand and four hundredths of an Earth day (1/86,400). Over time, however, it was determined that this definition was imprecise and not able to keep pace with the progress of science and technology. Small variations in the Earth's velocity of rotation made this standard inadequate. In 1960, the General Conference on Weights and Measures approved a new definition of the second, based on the Earth's annual orbit around the Sun. This new definition was much more accurate than the previous one, but also more practical. The second was defined as 1/31,556,925.9747 of a tropical year, or rather the period between the summer solstice of 1900 and the following one. Once again, however, just when it was thought that a fixed value had been established, scientific discoveries leapt forward and made the definition obsolete by bringing to fruition the practical application of the visionary idea launched by James Clerk Maxwell decades before.

In 1879, the father of electromagnetism wrote to his colleague William Thomson, suggesting that the period of vibration of a quartz crystal would be a more accurate measure of time than the period of the Earth's rotation, something that, as we have seen, came to be in the early decades of the twentieth century. Yet his imagination went beyond that, and he wrote that the measurement of time based on the oscillation of quartz depended on a piece of material and was, therefore, corruptible. A much better basis would be an invariable natural oscillation such as one tied to a property of the atom.

Maxwell and Thomson's imagined solution began to become reality in the 1940s, with the development of the first atomic clocks. Once again, the measurement is based on oscillating phenomena, events that repeat periodically with an identical duration and so can be counted, thus permitting the measurement of time. Unlike in the past, however, the event is no longer tied to an artifact like the pendulum or to a piece of quartz but is attached to a property of the fundamental element of matter: the atom.

We have seen in the chapter on the meter that one of the four basic postulates of Niels Bohr's atomic theory asserts that an atom emits electromagnetic radiation with a well-defined energy level when its electron, initially moving in a certain orbit, jumps into another orbit at a lower energy level. The frequency of the emitted radiation is equal to the difference between the two energy levels divided by Planck's constant. Conversely, to make an electron jump from one stable orbit to another with higher energy, it needs to be provided with an amount of energy exactly equal to the difference in energy between the two orbits. Because of the transitions of its electrons, every atom can therefore emit electromagnetic radiation with a limited set of preestablished levels of energy, or if you will, frequencies. As we have seen, this set of energy levels constitutes the spectrum of an atom, and each component of the periodic table has its own, different from the others. Moreover, the value of each frequency is totally independent of any human action.

Modern atomic clocks use one of the transitions of the cesium atom, specifically, the one that emits and absorbs packets of energy at a frequency of 9,192,631,770 oscillations per minute. This is a constant, immutable, and universal value, which serves as the basis for the modern definition of the second. The radiation emitted by cesium is measured and converted into an interval of time. Since 1967, the year of the 13th General Conference on Weights and Measures, the second has therefore been defined as the time needed to count 9,192,631,770 oscillations of the energy emitted by the cesium atom in a transition of its electrons between two specific energy levels. Just to be precise—and begging your pardon for being so fastidious—the chosen transition was the unperturbed ground-state hyperfine transition frequency of the cesium-133 atom, fixed and indicated by the symbol Δv_{Cs} in the new definition of the international system in the 21st edition of the General Conference on Weights and Measures in 2018.

The first atomic clock was built in the laboratory of the American National Bureau of Standards (today the National Institute of Standards and Technology, NIST), in 1949. Today, the atomic clock used at NIST to give the official time for the United States is so accurate that it can, at worst, lose or gain one second in . . . 300 million years.

There are certainly no more excuses for being late.

For Everyone and for Every One

From inaccuracies of dozens of minutes per day to those of a second every 300 million years, there can be no doubt that, over the past seven centuries, some significant progress has been made in the measurement of time. Yet, paradoxically, the more humanity tried to bridle and control time with ever more accurate measurements, the more the concept of time itself became elusive. Even

today, theoretical physics is interrogating itself about the meaning of time.

As for its measurement, an enormous paradigm change for the

definition of time came about in the seventeenth century. For over a millennium, the scientific and philosophical discourse around time was substantially at a standstill (recall Augustine of Hippo, who lived between the fourth and fifth centuries CE). When Copernicus published his first works and the scientific revolution took its first steps, the common understanding of time was still Aristotle's idea that time flowed only when something happened.

Think for minute of how time is kept in a basketball game. The time runs only when the ball is in play, when there is movement. If the game is interrupted for a foul, the clock stops. The comparison may seem flippant, but time in Aristotle's model is pretty much the same, something that runs only when events happen, when there is movement. Therefore, it is not absolute, just as the time effectively played in basketball games is variable. This was the dominant doctrine of time for over a millennium, and it pervaded medieval culture. We can imagine, therefore, the huge disruption that was set in motion by the advent of the scientific revolution, and especially by Galileo and later by Isaac Newton.

With Galileo and Newton, time takes on universal value, the same for everybody, everywhere. Time becomes an absolute parameter that serves as an index for natural systems, for the unfolding of physical processes. There exists, therefore, the possibility of establishing a universal clock with which to synchronize all other clocks. Time exists without regard to its perception.

Underlying this revolution in the conception of time are the principles of Galilean invariance, or Galilean relativity, which, as we have seen, propose that the laws of physics always have the same form in frames of reference that move at constant velocity in relation to each other. Or, in other words, where physics experiments cannot determine whether a body is standing still or moving at constant velocity.

Galilean relativity and the later work of Newton that codifies classical mechanics have had enormous and universal influence for centuries. With them, for example, the motion of the planets is described, as is that of the parts of a motor, and airplanes are designed.

The Galilean transformation also says something else: that time remains the same in the two frames of reference. Time is absolute. Newton imagined a space that was empty, infinite, where time passes even if everything remains empty and nothing happens. ("Absolute, true and mathematical time, of itself, and from its own nature, flows equably without regard to anything external, and by another name is called duration.") Returning to the world of sports, Newton's time is similar to time in a soccer game—ninety minutes that keep running even when the action is stopped. There is a precise line, the same for everyone, and that means every one, that separates the present from the past. My now is now for everyone. If I let this book drop, I can measure exactly how much time it will take to fall to the ground. And this interval of time is the same for everyone, wherever they happen to be.

Liquefying Time

Legend has it that, at the end of an interview with a by-then elderly Pablo Picasso, a journalist asked the master if he could have a sketch as a souvenir of their conversation. Picasso grabbed a pencil and a notebook and made a drawing. The journalist asked him, "Do you realize that it took you only a few seconds to make this sketch, and now I could sell it for thousands of pounds?" Picasso responded, "It didn't take eight seconds to make this drawing, it took eighty years."

It took Michelangelo four years to paint the Sistine Chapel, but other famous paintings took much less time. Salvador Dalí himself recounted that, to paint his famous painting *The Persistence of*

Memory, it took him only a couple of hours, the time it took for his wife, Gala, to go to the cinema to see a movie that he had skipped because of a headache.

That painting, from 1931, depicts a landscape on the Costa Brava featuring some melted, almost liquefied, pocket watches. It is a reflection on time, the ascetic time tracked by the watches but also by human experience. With the melting of the watches, objective time becomes flexible, subjective, personal. Relative. It is not surprising that many critics have suggested that Dalí was strongly influenced by Einstein's theory of relativity, which was much talked about at the time. Recall the fanfare surrounding the experimental confirmation of the theory achieved by the British astronomer Arthur Eddington, which we will address in the chapter on the kilogram. Having escaped from the academy, relativity became a topic of conversation and a leading player in cultural debate. In 1929, just two years before Dalí's painting, the *New York Times* attributed to Einstein the following remark: "When you sit with a nice girl for two hours you think it's only a minute. But when you sit on a hot stove for a minute you think it's two hours. That's relativity."

We don't know if Einstein was really the author of this quip (which, obviously, would be equally valid even if the roles were reversed or if the protagonists were of the same gender). It is certainly true, however, that his theory of relativity revolutionized the concept of time by denying its absoluteness. Time is no longer an absolute concept. It flows in different ways in frames of reference that move at different velocities. Two events that are simultaneous for an immobile observer may not be for an observer in motion. Less than three centuries after Galileo, time experienced another revolution.

Up until the end of the eighteenth century, in fact, the basic elements of mechanics were the Galilean principle of relativity and the concept of absolute time. Then came the development of electromagnetism, the science that described the electric and magnetic phenomena that were increasingly a part of society, of the economy, of daily life. Just think of the revolutionary developments of electric light, Marconi's first transoceanic transmissions, the first electric motors. In 1892, between Rome and Tivoli, in the nearby Sabine Hills, the first experimental line for the transmission of electricity went into active service. In short, the practical application of electromagnetism was in full swing. So it was a real problem when physicists noticed that the equations that describe it, Maxwell's laws, were not consistent with Galilean relativity. Maxwell's equations are not invariant under Galilean transformation. Electric and magnetic phenomena in a moving frame of reference are different from those in a stationary one. A real problem!

A problem rooted in a simple fact. Do you remember Superman? "Faster than the speed of light!" . . . Well, as we have seen in the preceding chapter, that is a flight of fancy.

Relative Present

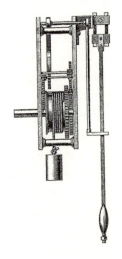

"Tell me, what is the use of these experiments of yours?" Supposedly, one day the English chancellor of the exchequer asked this question of Michael Faraday. An insidious question, given that in the mid-nineteenth century, Faraday was a scientist working in Her Majesty's service and was therefore rather sensitive to the finance minister's opinion of his work. Faraday, however, was not cowed: "I can't tell you exactly, but one day you can tax it." And he was right. What he was working on was the experiment that would demonstrate the possibility of transforming mechanical energy into electrical energy thanks to the movement of a conductor in a magnetic field. In essence, the prototype of modern generators of electricity.

A glance at the taxes on your electric bill confirms that Faraday had a good imagination.

Those were years of great ferment for electromagnetism, which on the one hand was finding more and more practical applications, while on the other hand was being given a complete theoretical apparatus by Maxwell's equations, developed in part based on the work of Faraday. However, Maxwell's equations, which describe the behavior of the electric and magnetic field, have a precise consequence. Think of a light source, a light bulb, for example. The light it emits always moves at the same constant velocity, 299,792,458 meters per second, regardless of the velocity of the source that emits it. In other words, as fast as we might be moving, light is always faster than us by exactly the same amount: 299,792,458 meters per second. Light always moves at the same speed in whatever frame of reference.

This was a big problem for Galilean invariance. Einstein gets the credit for the solution. He started from two givens:

1. All the laws of physics remain the same in all frames of reference that move at constant velocity with respect to each other. In other words, it cannot be determined through a law of physics if a frame is in uniform motion, or in other words its absolute velocity.
2. The speed of light is the same in all frames of reference in motion.

To satisfy his conditions, Einstein modified the Galilean transformations between two frames of reference in motion with respect to each other with velocity v, which became:

$$x' = \frac{x - vt}{\sqrt{1 - \frac{v^2}{c^2}}}$$

$$y' = y,$$

$$z' = z, \qquad (1)$$

$$t' = \frac{\left(t - \frac{vx}{c^2}\right)}{\sqrt{1 - \frac{v^2}{c^2}}}$$

Perhaps a bit more complicated, but note them well: they are revolutionary!

In the Galilean transformations, time is a parameter independent of frames of reference and position. Time flows equally in all frames of reference. With Einstein, time loses this prerogative of absolute identity, its status as invariable. Time t blends with space (described by the coordinates x, y, z); it becomes relative. Time is no longer an absolute, independent quantity. The theory of relativity tells us that time flows more slowly in systems in motion: it dilates. If the clock in a hypothetical train traveling at a velocity proximate to the speed of light shows the passing of one second, the watch of the observer standing in the station will show a time that is longer than one second. For example, if the train is traveling at 270,000 kilometers per second, that is, at nine-tenths of the speed of light, when ten minutes have gone by for the observer in the station, the time elapsed for the passenger will be just over four minutes. If you want to stay young, find a train that travels at close to the speed of light.

Einstein's relativity also upset the concept of absolute simultaneity, which was implicit in the Galilean transformations. The concepts of "now" and "in this instant" no longer have a universal meaning. While for the prerelativity way of thinking, the concept of "now" is taken for granted and the expression "right this minute" has a precise meaning that is valid any place in the universe, for Einstein the simultaneity of two events in a given frame of reference may no longer be the same in a different frame. Two events that occur in different places but simultaneously—which are recorded, that is, at what is the same instant for an immobile observer—are no longer simultaneous when seen in a frame of reference in motion with respect to the station: on the train, for example.

We can no longer speak of "now." Let's take this precise instant, in which you are reading this line. Before, with Newtonian relativity, there existed a clearly defined line of demarcation, which extended throughout the entire universe, between past and future with respect to my "now," to my "right this minute," to the precise moment in which you finish reading this word. An absolute border.

With Einstein, everything changes.

The Shadow of Space-Time

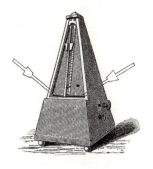

Let's suppose that astronauts have gone on an ambitious space mission and have arrived on Proxima Centauri, one of the stars closest to Earth. In fact, this star is about four light-years away from our planet, about 40 trillion kilometers. This means that light takes four years to travel to Proxima Centauri from Earth, and that is also the minimum time that a signal takes to travel from that star to us.

On landing, the astronauts open a chat on some social network— yes, they too have friends they want to impress—and during the chat they start transmitting a live (for them) video to show us how things are going. Those images will reach us four years after they send them. The whole telecast of their mission will be delayed by four years.

"Here, now you can see some inhabitants of Proxima Centauri moving toward me," one of the astronauts tells us in the video, zooming in on the natives. But the astronaut's "now" has a meaning very different from our "now."

Our "now," which reaches us by way of the telecast, refers to four years ago on the star. A present, a universal "now" that separates past and future, does not exist. We have no idea what is happening on Proxima Centauri in this moment. We cannot know now if the natives offered the astronauts some coffee or if the astronauts less amiably transformed the natives into fuel for their spaceship. We will find out in four years. And if you are reading this book by the light of the Sun, you do not know whether our little private star is still shining now. The Sun could actually have gone dark, and we would not realize it until eight minutes later, which is how much time it takes for its light to reach us here on Earth. The James Webb Space Telescope, which recently started producing its beautiful images thanks to a partnership among NASA, the European Space Agency, and the Canadian Space Agency, exploits this physics process to explore the origin of our universe. The light measured today was in fact emitted by galaxies in the early universe. As the Webb website says, "Webb will directly observe a part of space and time never seen before. Webb will gaze into the epoch when the very first stars and galaxies formed, over 13.5 billion years ago."

"Now" is not something that is physically observable, because in the case of Proxima Centauri it takes us, here on Earth, four years to measure it. On that star, there are a series of events that have already happened—which without doubt belong to our past—and that are before four years ago. Then there are other events, which will belong to our future—and which will happen in no less than four years. Yet there is also an undefined period of eight years, which belongs neither to our past nor to our future. Eight years of events, which, on the one hand, we do not yet know about and which, on the other, we cannot influence. If our supercomputer were to predict that six years from now it will rain on Proxima Centauri, we could send a message to our astronauts and influence the future (they could buy umbrellas and avoid getting wet). But if we predict that it will rain only three years from now, there is nothing we can do to warn them.

The past is that set of events that can send light signals to observers and thus influence them. The future is that set of events to which observers can send signals of light and that, in principle, can then be influenced by the observers themselves. Then there is a new series of events in space-time that we cannot influence now from where we are, and that cannot influence us here and now, because nothing can travel faster than light. This new set of events represents a sort of extended present, neither past nor future, a consequence of Einsteinian relativity. The duration of this extended present depends on position: sixteen minutes in the case of the Sun, eight years in the case of Proxima Centauri. Whereas before Einstein, space and time were clearly distinct entities, now they blend. They must inevitably be considered together and become space-time. It is not something that is easy to accept. The notion of absolute time is deeply rooted in our experience.

In the words of Hermann Minkowski, a mathematician who was one of Einstein's professors in Zurich, "Henceforth space by itself, and time by itself, are doomed to fade away into mere shadows, and only a kind of union of the two will preserve an independent reality, space-time."

Everyday Relativity

Time is also influenced by the mass of objects in the vicinity. This is the result of the general theory of relativity, $$E=mc^2$$ the extension and completion of the special theory of relativity.

With his general relativity, Einstein put together his principle of relativity with Newton's universal law of gravity, another of the fundamental laws of physics, which describes how two masses interact with each other by way of an attractive force. Space-time is enriched by gravity, this force at a distance that controls the movement of the planets. Space-time is no longer something empty, rigid, but becomes a flexible entity, a sort of network whose webbing intersects with the lines along which the force of gravity operates. A network that bends in proportion to objects with mass—the greater the mass the greater the bend—just as a mattress forms a hollow equivalent to the size of the person who sits on it.

Massive objects bend space, and the curvature attracts other bodies toward them. The Earth revolves around the Sun, thanks to its velocity, just as a cyclist does in a track race. Can you envision those inclined tracks they use in the races that are televised during the Olympics? The cyclists remain up high on the track only as long as they are racing. When they stop racing, they inevitably drift down toward the bottom of the track, like coins tossed into those funnel-type things often seen in museums. If it slowed down, the Earth would be attracted by and crash into the Sun. The general theory of relativity helps us understand, for example, black holes, extremely "massive" objects that attract toward themselves everything around them, including light, which then never manages to come back out.

The general theory of relativity also contributes to the further modification of our concept of time. Indeed, time is modified not only by movement but also by gravity, by the presence of masses. Time loses another part of its "absoluteness"—if there was any left after special relativity—and now it flows at different speeds depending on the masses that are present in the vicinity.

Time passes more slowly near to a mass, and therefore, on Earth it goes by faster up high—more distant from the center of the Earth—than down low. As if to say that tall people age faster than short people. On a human scale, the effects are extremely small but measurable, as demonstrated by the experiment of Hafele and Keating, mentioned at the beginning of this chapter. As recently as 2018, a transportable atomic clock was taken to a laboratory located in the Fréjus Road Tunnel (at an altitude of about 1,200 meters) in the

Alps by researchers from the Italian National Meteorological Research Institute, who verified that time flowed faster on the mountain compared to their headquarters in Turin, situated at an altitude of about 200 meters.

One of the most striking experimental demonstrations of general relativity is the recent measurement of gravitational waves. These are very weak wrinkles in space-time, generated by modifications in the distribution of masses on a cosmic scale, such as the collision between two black holes. These wrinkles extend outward like waves in the ocean. But the modification of time described by the theory of relativity also has a much more practical application, one that we even carry in our pockets. By now, all of our cell phones have an incorporated GPS (Global Positioning System). Well, if the GPS did not correct for the modifications in the flow of time owing to relativity, it would make kilometric errors in the calculation of positions.

A GPS satellite orbits at about 20,000 kilometers from the surface of the Earth, and it moves at a velocity of about 14,000 kilometers per hour with respect to Earth. If we run the numbers, we find that just the effect of special relativity would cause the clocks of the GPS to slow down by about seven microseconds per day. To be sure, we are talking about a few millionths of a second. Yet if the GPS clock did not take this into account, since the electromagnetic waves with which the satellites send signals to our receivers on Earth cover about 30 centimeters in a nanosecond, seven microseconds would correspond to an error of two kilometers! Moreover, if we add in the effect of gravity, given the distance of 20,000 kilometers between the satellites and the receivers, the positioning error would increase to 18 kilometers! In other words, if it hadn't been for Einstein, we would never have found that marvelous country house nestled sweetly in the hills of . . . .

THREE

The Kilogram

Letters

Dear Edoardo,

Mein Führer!

It is hard to imagine that two letters with these salutations, one written by hand to a friend, the other typewritten and sent to Adolf Hitler, might have something in common. Yet there is actually a lot that unites the two missives. For starters, the dates they were written, the first on August 15 and the second on October 25, just a few weeks apart in the year 1944. Then the anxiety for the fate of loved ones—a father-in-law and a son—the passion for the writers' own work that can be read between the lines, and the drama of a time oppressed by horrible dictatorships, fascism, and Nazism that would do away with both of those loved ones. Above all, that the authors of the letters were two of the most famous physicists of all time, Enrico Fermi and Max Planck.

The summer of 1944 was a period of change for Enrico Fermi. The letter was written in Chicago, but Fermi was about to leave, on his way to Los Alamos, New Mexico. A victim of Benito Mussolini's racial laws (Fermi's wife, Laura, was Jewish), he had left Italy in 1938, when he was awarded the Nobel Prize. He went to Stockholm to receive it, and from there, after a brief stop in Copenhagen for a visit with Niels Bohr, he embarked for the United States. His American period began at Columbia University, in New York, and then he moved on to the University of Chicago. There, in 1942, Fermi built the first atomic pile and produced the first fission chain reaction, an experiment that threw open the doors to the exploitation of nuclear energy. Then, in 1944, he was called to Los Alamos by Robert Oppenheimer to work on the Manhattan Project, which would produce the first American atomic bombs, later dropped on Hiroshima and Nagasaki.

The addressee of Fermi's letter was Edoardo Amaldi, one of the youngest of the group of scientists in Rome known as the Via Panisperna Boys. Rome had just been liberated. American troops, under the command of Gen. Mark Wayne Clark, had entered the city on June 4, and Fermi took advantage of the reopening of communications to write to his colleague and friend. "Dear Edoardo, I had some recent news of you from Fubini[1] on his return from Italy. Now that postal communications with Rome have been officially reopened, I hope this letter has a good chance of reaching you."

---

1. Eugenio Fubini was an Italian physicist who studied under Fermi in Rome and taught at the University of Turin until being forced to leave Italy in 1938 by the racial laws. In the United States he went on to become assistant secretary of defense in the Kennedy administration and then vice president and chief scientist at IBM before founding his own consulting firm.

Then he immediately writes about Augusto, father of his wife, Laura ("Lalla") Capon. "As you can imagine, Lalla has been very upset by the news about her father. The uncertainty about his fate is much worse than knowing he is dead." Augusto Capon, a Jew, was a prominent admiral in the Italian Royal Navy and a friend of Mussolini. Until 1938, he was the head of the navy's Secret Information Service. That was not enough to save him when, on October 16, 1943, Italian and German soldiers combed the city searching for Jews. That very day, Capon wrote in his diary, "Incredible things are happening in Rome: this morning groups of fascists, they say together with some German soldiers, have picked up Jews of any age and gender and taken them to some unknown place. That this happened is certain, how is not." Capon died in Auschwitz the following week.

Later in the letter, Fermi's passionate concern for the fate of physics in his country comes to the fore. After years of darkness and the dissolution of Italian research, Fermi hints at a note of optimism. "I was very pleased to hear that you and Wick[2] are hoping you will soon be able to get back to your scientific work, and that you are looking to the future with a certain degree of optimism. Judging the situation from this side of the Atlantic, I sometimes hope that the reconstruction of Italy may be less difficult than that of other European countries. Certainly fascism fell in such a miserable fashion that it doesn't seem possible it has left any regrets."

It is to Adolf Hitler, who was at the center of the horrors of the war and the extermination of the Jews, that Max Planck, the eminent German physicist, winner of the Nobel Prize in 1918, and one of the fathers of quantum mechanics, addresses his letter. He does so to plead for mercy for his son Erwin.

Planck had already met Hitler personally in 1933, just after his

2. Gian Carlo Wick was an Italian theoretical physicist who was Fermi's assistant in Rome and after 1946 a professor at several American universities.

rise to power. At the time, the 75-year-old Planck was probably the most authoritative scientific personality in Germany and president of the prestigious Kaiser Wilhelm Society for the Advancement of Science, which led German research. In that capacity, Planck requested a formal audience with the new chancellor, who had taken office just a few months before. The purpose of the audience was for him to pay his respects, but Planck—who, unlike many of his colleagues, was never to leave Germany, remaining loyal to his country despite his disagreement with the madness of Nazi policies—took advantage of the occasion to ask the Führer's clemency on behalf of Jewish scientists. In those very months, Planck's Jewish colleagues were beginning to suffer the initial consequences of the racial laws and were being dismissed from their positions. Among them was Planck's friend Fritz Haber, winner of the Nobel Prize in Chemistry in 1918, who had also distinguished himself (so to speak) as the mastermind of chemical weapons used in World War I. Whether out of conviction, lack of courage, or just plain realism, Planck did not express to Hitler his disapproval of the racial laws as such. If he had done so, he probably wouldn't have made it back home. Yet he did try pragmatically to convince him that it would be self-destructive for Germany to deprive itself of the talents of its many Jewish intellectuals. He recalled how, without Haber's patriotic (and terrible) scientific contribution, Germany would probably have been defeated very early on in that first world war and that many eminent German scientists were Jews.

Hitler wanted nothing of it. "I have nothing against Jews in themselves. But the Jews are all Communists, and it is the latter who are my enemies; it is against them that I am fighting," he responded to Planck, mocking him—it seems—with a flat "I guess that means we'll do without science for a few years." Considering the number and the caliber of the physicist fugitives from Nazi-fascism who contributed to the development of the atomic bomb, this was certainly not a renunciation without consequences. The conversation quickly degenerated into a rabidly irrational monologue, in the face of which there was nothing for Planck to do but remain silent and withdraw.

When Planck tried to contact the Führer a second time, this time in writing, 11 years had passed since their first encounter. Hitler had dragged the world into war, and by now, the tide had turned against Nazi Germany. Planck was old and wearied by life. His scientific success had been accompanied by continual family dramas. In 1909, he had lost his first wife. His firstborn son, Karl, was killed in the Battle of Verdun during World War I. Two twin daughters, Grete and Emma, both died in childbirth, in 1917 and 1919, respectively. In 1944, his home in Berlin was destroyed in a bombardment. His second son, Erwin, was taken prisoner in 1914 but had managed to return home. After the war, he occupied various positions in government, rising to the level of secretary of state under Chancellors Franz von Papen and Kurt von Schleicher. When the latter resigned in 1933 and Hitler took power, Erwin resigned his office and devoted himself to business, all the while maintaining a strong interest in politics with views increasingly critical of Hitler.

In the closing months of 1943, Erwin joined with the conspirators who were plotting Operation Valkyrie, a coup attempt to overthrow Hitler and negotiate a peace treaty with the Allies. The brain behind the operation was army colonel Claus von Stauffenberg. The plan was to assassinate Hitler with a bomb inside his general headquarters, known as the Wolf's Lair, in Rastenburg (today Kętrzyn, Poland). On July 20, 1944, the bomb was placed in a briefcase and left by von Stauffenberg under the table in the conference room, near Hitler, during a meeting of the general staff. Owing to a series of coincidences—only one of the detonators worked and an officer happened to move the briefcase with his foot just before the explosion—the bomb exploded and caused a lot of damage, but Hitler was only slightly wounded. Soon afterward, the conspirators and thousands of other people were arrested. Among them was Erwin Planck, picked up by the Gestapo and condemned to death forthwith.

His octogenarian father tried desperately to save his son from the gallows by exploiting his contacts and his fame as a great scientist. On October 25, he wrote to Hitler:

> My Führer!
>
> I am most deeply shaken by the message that my son Erwin has been sentenced to death by the People's Court.
>
> The recognition of my achievements in service of our fatherland, which you, my Führer, have expressed towards me repeatedly and in the most honoring way, makes me confident that you will lend your ear to an imploring 87-year-old.
>
> As the gratitude of the German people for my life's work, which has become an everlasting intellectual patrimony of Germany, I am pleading for my son's life.
>
> Max Planck

In this case, too, as with Fermi's letter, the passion for physics stands out. Planck proudly recalls his contribution, which increasingly would become the heritage not only of his homeland but of all humankind. He, Nobel Prize winner, loyal to his country right to the end, begs for pity. An act of desperation, and perhaps, as he is writing, Planck remembers his encounter of 11 years before, and he knows that, just as Hitler had no pity for the Jews then, he would not have pity for Planck's family now. Could science, even for a minute, dent the delirium of Nazi fanaticism?

Erwin Planck was hanged on January 23, 1945. Four days later, the Red Army liberated the extermination camp at Auschwitz and the world discovered the horror of the Shoah.

Talents and Carob Seeds

For it is like a man going on a journey, who summoned his slaves and entrusted his property to them. To one he gave five talents, to another two, and to another, one, each according to his ability.

Then he went on his journey. The one who had received five talents went off right away and put his talents to work and gained five more. In the same way, the one who had two gained two more. However, the one who had received one talent went out, dug a hole in the ground, and hid his master's money in it. After a long time, the master of those slaves came and settled his accounts with them.

The parable of talents, taken from the Gospel according to Matthew, is probably one of the best-known passages from the New Testament. The talents entrusted by the master to his servants represent God's gifts to humankind, and those who render them fruitful are rewarded. Today, in the wake of the Bible story, the noun *talent* is typically associated with its metaphorical meaning, but it is intriguing to find out that the talent described by Matthew the evangelist was actually quite . . . solid. The talent, in fact, was a unit of measurement for weight used as early as the Mesopotamian civilizations. In ancient Greece, the talent was equal to 26 kilograms, the weight of the amount of water needed to fill a certain type of amphora. It could also correspond to an equal weight of a precious metal. The talent in the parable was most likely made of silver, which is thought to have been sufficient to pay the monthly salary of the entire crew of a trireme, about 200 people.

As with the measurement of length and time, humanity has been concerned with the measurement of the weight of bodies since the dawn of civilization. To be precise, rather than weight, it would be better to speak of mass, which in effect is the physical quantity that appears in the international system. In ordinary language, the two terms *weight* and *mass* are often confused, but only because we live on the surface of the Earth and mass is measured on common scales thanks to the gravitational attractive force (the weight force) that our planet exercises on any object and that keeps us solidly anchored on it. As Newton understood—and as we shall see in detail as this chapter unfolds—the weight force on an object on the terrestrial surface is directly proportionate to its mass. Measuring a force is relatively easy, especially if done in a comparative way, as in the two-pan balance—the one that you see in the halls of justice and that was engraved on the face of a commemorative U.S. silver half-dollar coin issued in 1936—or in the lever scale used by greengrocers. In both cases, the mass of the object to be measured is compared to a standard mass, typically a kilogram or a submultiple, thus comparing the two weight forces.

As with the measures of length and time, the primary motivation for codifying the measure of mass came from the demands of everyday life, particularly commerce. Ancient archaeological finds related to the use of scales come from sites in the Indus Valley, today in Pakistan, and date from 2400 and 1700 BCE. About coeval are the Egyptian finds (ca. 1878–1842 BCE), although it is highly likely that the use of weighing goes much further back, considering the spread of commerce in the civilization that developed along the Nile. Don't forget that, even then, the scale already had a metaphorical meaning. The god Anubis, protector of tombs and of the dead, represented with a jackal's head, used a two-pan balance to weigh the heart of the deceased and to compare it to a feather. The measurement would determine the deceased's admission to the next world. Typically, the archaeological finds are standard weights, polished stones that were used to balance the mass to be measured, or the arms of the balance, a noun that derives etymologically from the Latin *bis* (twice) plus *lanx* (pan).

The primary unit of measurement for mass in ancient Rome was the *librae* (pound), which takes its name from the Latin *libra* (scale). Many ancient units of measurement have their origins in seeds of grain or carob. The latter is the origin of the carat, still employed to measure the weight of precious stones or gems and used in the past to identify the twenty-four quotas in which the cargos of Venetian or Genoese merchantman galleys were divided. The twelfth-century *Tractatus de ponderibus et mensuris* says, for example, "Per

Ordinance of the whole realm of England the measure of the King is composed namely of a penny, which is called a sterling, round & without clipping, weighs thirty-two grains of wheat in the middle of the Ear. And an ounce weighs twenty pence. And twelve ounces make a pound." The nexus between measures of weight and commerce is also confirmed by the fact that most monetary denominations derive from measures of weight, as the Italian historian Alessandro Magno Marzo notes in his book *L'invenzione dei soldi* (The Invention of Money): lira, pound, peseta, and mark, for example.

Little wonder, therefore, that, as happened for the unit of measurement of length, the decisive push toward the definition of a universal system for measuring mass came from the growing globalization of commerce and the spirit of the Enlightenment that swept across Europe in the eighteenth century.

One illustrious exponent of that culture was Giovanni Fabbroni, born in Florence in 1752. Chemist, naturalist, agronomist, and economist, he was the director of both the Royal Museum of Physics and Natural History in Florence and the mint of Pietro Leopoldo, grand duke of Tuscany. A man of many interests, Fabbroni was also noticed by Thomas Jefferson, who a few years later would become the third president of the United States.

Together with the French scientist Louis Lefèvre-Gineau, Fabbroni played a crucial role in the definition of the kilogram. Thanks to them, in 1795, during the French Revolution, the kilogram was defined as the weight of a cubic decimeter of water at a temperature of 4°C, in contrast to its definition a few years earlier as the weight of a cubic decimeter of water at 0°C. That difference of four degrees was very important: Fabbroni and Lefèvre-Gineau realized that using water at 4°C—at its maximum density—would permit a much more stable definition of the kilogram.

Incidentally, this behavior of water is peculiar, different from that of most known substances, and crucial for the ecosystem. It is the reason that, in winter, bodies of water freeze starting from the surface, but not at their deeper levels, which remain in the liquid state and assure the survival of aquatic species. In 1799, following the definition of the kilogram by Fabbroni and Lefèvre-Gineau, the prototype of the kilogram was made. Known as the *kilogramme des archives*, it was a cylinder of platinum with a mass identical to that of a cubic decimeter of water at 4°C, conserved in the National Archives in Paris.

The Metre Convention in 1875 confirmed the definition of the kilogram and materialized it in a new artifact: the International Prototype Kilogram (IPK), a cylinder made of 90 percent platinum and 10 percent iridium, conserved at the International Bureau of Weights and Measures. Numerous copies of this prototype were made; six were kept at the bureau and the others were distributed to the signatory countries. The United States received copies number 4 and 20 in 1890, which were joined in 1996 by copy number 79.

As in the case of the meter, however, reliance on material objects, no matter how refined, ran the risk of the prototype's deterioration over time, owing to contamination or corrosion, which indeed occurred. A series of measurements, initiated in 1899 and continued until 2014, verified that the six sister exemplars of the IPK and some of the national copies were gaining weight with respect to the kilogram of reference. The copies gained an average of 50 micrograms in 100 years. Though nothing compared to the nightmares of someone on a diet looking with ravenous eyes at a plate of pasta with tomato sauce, nevertheless, it was enough to put into doubt the rigorous accuracy required of a universal standard, all the more so in the face of the increasingly pressing demands for accuracy on the part of modern science and technology.

Guess Who's Coming on Our Honeymoon?

Arthur Eddington was a great British scientist who lived at the turn of the nineteenth century. Astronomer and physicist, he was the

author of pioneering studies on the behavior of the stars. He was the first, for
example, to hypothesize that nuclear
fusion is a fundamental process in the
dynamics of stellar energy. Eddington
was also a great admirer of Einstein,
and he tried to overcome the isolation
of German-language scientists during
and right after World War I by publi-

cizing the general theory of relativity in the English-speaking world.
Yet he didn't stop there. He was also the first to demonstrate the
theory experimentally, using the mass of the Sun. Our private star,
which from the Earth looks like a small disc—a little bigger at sun-
set, when it is lower on the horizon—is actually a rather massive
body. Indeed, its mass described in kilograms is a number with 31
figures, about 330,000 times bigger than the mass of the Earth.

The general theory of relativity was revolutionary and mathematically complicated, and it was not accepted by everybody right
away. Einstein himself was aware of the need for an experimental demonstration and suggested measuring the deflection of light
coming from the stars caused by the mass of the Sun, as predicted
by his theory. The brightness of the Sun, however, made direct
observation impossible, and it was certainly not possible to turn it
off: except for—and this is where intuition comes in—during a
total eclipse, which would allow someone to photograph the stars
near the Sun and to verify if there was an apparent shifting of their
position compared to when they were observed in proximity to
the Sun.

The first to take up the challenge was Erwin Freundlich, an en-
thusiastic astronomer in Berlin. Freundlich had scheduled his wed-
ding for the summer of 1913 and planned his honeymoon in the
Swiss Alps so that he could meet Einstein in Zurich and talk about
his experiment. We do not have any direct evidence about the reac-

tion of Mrs. Freundlich. The meeting did take place, however, and gave rise to a plan for an expedition to Crimea, led by Freundlich, on the occasion of the eclipse forecast to take place on August 21, 1914. But Freundlich was unlucky. Just when he arrived in Crimea, World War I broke out in Europe. On August 1, Germany declared war on Russia. When the Russians stopped Freundlich on their soil—a scientist from an enemy power armed with binoculars and telescopes—they were not inclined to believe that he was there to measure the deflection of starlight. He was arrested and his equipment seized. One month later, he was freed thanks to a prisoner exchange.

The baton was passed to Eddington, who proposed to take advantage of the eclipse of May 29, 1919. Given the times, it was not such a simple idea. Great Britain and Germany were still shedding each other's blood, and for the British there was little appeal in organizing an expedition to demonstrate the validity of a theory produced by a German scientist. Nevertheless, Eddington managed to pull it off, and as he later wrote, "By testing the 'enemy' theory our national observatory kept alive the finest traditions of science and the lesson is perhaps still needed today."

To observe the eclipse, Eddington's team split into two groups. Eddington and a few colleagues went to the island of Principe, off the west coast of Africa. The others went to Sobral, Brazil. That day on Principe, the sky was cloudy and the bad weather could have sent months of preparation up in smoke, but Eddington, unlike Freundlich, was a lucky guy. Just as the eclipse was about to begin, the blanket of clouds opened up. This allowed him to photograph some of the stars in the cluster of the Hyades. When the measurements were interpreted, they confirmed the theory of general relativity. On November 6, 1919, the results were presented to the Royal Astronomical Society. The news—until then circumscribed within the narrow circle of physicists—immediately bounced from one corner of the world to the other. Einstein's notoriety went global. The *Times* of London proclaimed "REVOLUTION IN SCIENCE / NEW THEORY OF THE UNIVERSE / NEWTONIAN IDEAS OVERTHROWN." The headline in the *New York Times* was slightly more sensationalistic: "Lights All Askew in the Heavens: Einstein's Theory Triumphs." Partial justification for the American daily may be that, since they had no scientific correspondent in London, the story was reported by a golf correspondent. . . .

Three, Twelve, None?

Eddington was certainly not short on brilliance or good fortune. The same cannot be said for modesty. At least not if one lends credence to the anecdote of how, at a meeting of the Royal Society, a physicist approached Eddington and addressed him as one of the three persons in the world who understood the theory of relativity. When he failed to reply, the other physicist admonished Arthur not to be so shy, to which he replied, "Oh, no, I was just wondering who the third one might be!"

A sort of semiserious debate has developed around the number of people who understood the theory of relativity, leading the well-known online platform Quora to publish an exchange of questions and answers on the theme. One contribution to the debate came from Nobel Prize winner Richard Feynman, who, with the grace that distinguished so much of his scientific popularization, wrote in *The Character of Physical Law* (1965): "A lot of people understood the theory of relativity in some way or other, certainly more than twelve. On the other hand, I think I can safely say that nobody understands quantum mechanics."

Not coincidentally, Feynman mentions together the two theories that have revolutionized modern physics, and which more or less in the same time period were theorized and experimentally confirmed thanks to objects that are extremely light or extremely heavy, or anyway enormously far removed from the scale of weight we are used to in our daily lives. While the huge mass of the Sun proved the theory of relativity, it was the experimental investigation of the microscopic world and elementary particles, such as atoms, that opened the way to quantum physics. A hydrogen atom, for example, has a mass of one billionth of a billionth of a billionth of a kilogram. If, on one pan of a hypothetical balance, we were to put the Sun, then in order to level the other pan with only hydrogen atoms, the number of them that we'd need would have fifty-seven zeroes. Infinitesimal fractions or exorbitant multiples of the kilogram have thrown open the doors to new interpretations of nature, which are objectively not easy fully to understand. Feynman's claim sounds like an exaggeration, but read more deeply it is still true today.

First of all, there is no single founder of quantum mechanics (if there were, he or she, at least, should have fully understood it). As David Griffiths has written in his excellent *Introduction to Quantum Mechanics* (1995), "Unlike Newton's mechanics, or Maxwell's electrodynamics or Einstein's relativity, quantum mechanics was not created—or even definitively packaged—by one individual, and it retains to this day some of the scars of its exhilarating but traumatic youth." Though it works quite well, why it works and its deeper meaning are still subjects of study. "There is no general consensus," Griffiths continued, "as to what its fundamental principles are, how it should be taught, or what it really 'means.' Every competent physicist can 'do' quantum mechanics, but the stories we tell ourselves about what we are doing are as various as the stories of Scheherazade, and almost as implausible."

In other words, quantum mechanics says a lot of things that describe physical systems very well, but we still don't know what it means. There are a lot of theories on the interpretation of quantum mechanics that interrogate and challenge each other, sometimes overflowing into philosophy. These incursions attract a large audience, but by abandoning the solid ground of science, they adventure out into much more slippery terrain and risk giving an image of the theory that is less than fully faithful. Meanwhile quantum mechanics is a theory that, in terms of predictions, reproducible results, and descriptions of physical systems, is extremely robust. This includes the theory that originated it, and that we owe to none other than Max Planck.

The Light of the Black Body

It might seem paradoxical for such a complex theory that describes systems like elementary particles, apparently so distant from our daily lives, but the quantum revolution begins with the analysis of a phenomenon that is part of everyone's experience: thermal radiation. Surely everyone has observed that a hot piece of metal emits a light, which typically goes from reddish to white. Think about a stoker in the fireplace, or the old incandescent light bulbs, which were outlawed in Europe in 2009 and are due to be phased out in the United States after 2023. That light is composed of electromagnetic waves and is called thermal radiation because it is emitted by a body as a result of its temperature. We ourselves, given our body temperature of about 37°C (98.6°F), emit thermal radiation in the infrared, which can be seen using the right instruments. This is the basic principle relied on for the measurement of body temperature with the thermal scanner, which unfortunately is now familiar to everyone because of the Covid-19 pandemic.

Thermal radiation increases rapidly in relation to increases in temperature: the hotter the body, the more intense its temperature. This phenomenon is described by Stefan's law, according to which the energy radiated per second by unit area of a body at temperature T is directly proportionate to T^4. In other words, when the temperature of an object doubles, the power emitted becomes sixteen times greater.

The specific properties of thermal radiation depend on the composition of the object that emits it. There are some bodies, however, that emit thermal radiation with universal properties: black bodies. In name and in fact. A black body is a body that absorbs all incident radiation, and precisely for this reason, we see such bodies as black. In the same way, colored objects appear as such because they reflect part of the light, specifically the part that has the color that we observe. The fabric of your red sweater is not actually red; the red we see is the light reflected by the material.

With the advent of electromagnetism, toward the end of the nineteenth century, the first accurate measurements were made of the properties of black-body radiation and the first theories emerged. Between 1900 and 1905, Lord Rayleigh and James Jeans attempted to analyze experimental observations by applying the laws of classical physics. Despite rigorous work, they did not succeed in describing the experimental evidence. In hindsight, the reason is simple. Within the confines of classical physics, their work was perfectly correct. The problem is exactly that: in these newly discovered realms, classical physics is no longer valid.

It was the genius of Max Planck who realized this and planted the first seeds of quantum mechanics in 1900. Planck followed the theory of Rayleigh and Jeans but advanced a revolutionary hypothesis: the energy of electromagnetic waves cannot vary continuously but can only take on a set of values, which are multiples of a fundamental quantum:

$$E = hf, \tag{1}$$

where E is energy, f is the frequency of the wave, and h is a universal constant. In his honor, this would later be called Planck's constant, and as we shall see, it is the basis for the most recent definition of the kilogram. In classical physics, light was understood as an electromagnetic wave whose amplitude could vary continuously and could therefore assume any value. Now, light is instead understood as being absorbed and emitted in discreet packets similar to particles. It's like buying milk at the supermarket. We can buy one bottle, two, three, . . . but always a whole number. Hard to show up at the checkout line with 27.0895 gallons of milk.

With this law, Planck introduced the concept of the quantization of energy, which turns out to be essential for the description of a physical phenomenon. At the microscopic level, nature is discontinuous, and science now began to listen to it. For all that scientific thought that was still dominated by the principle *Natura non facit saltus* (Nature does not take leaps), this was an enormous paradigm change.

Quantum mechanics had arrived.

Even Nobels Make Mistakes

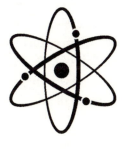

At the Royal Swedish Academy of Sciences in Stockholm they are demanding. To the point that they are willing to postpone the Nobel award ceremony if they are unable to find invitees they consider worthy. That is exactly what happened in 1921, when the selection committee for the Nobel in physics decided that none of the candidates was adequate to receive the prize. By the way, the list of the illustrious "rejects"—if you will—is now public. All candidates, even those who did not make it, are made public after 50 years. If you are thin-skinned, remember, being a Nobel candidate can have its disadvantages.

If the prize is not awarded one year, the rules provide that the committee can keep it on hold for the next year, when it will give two of them. Apparently, in 1922 the Swedish academy had no lack of candidates, since it awarded two prizes: the one for 1921 to Albert Einstein and the one for 1922 to Niels Bohr, the father of modern atomic theory and one of the founders of quantum mechanics.

At the awards ceremony on December 10, 1922, the two winners were presented to the king of Sweden by Svante Arrhenius, he, too, a Nobel winner for chemistry in 1903 and at the time president of the physics section of the selection committee. Arrhenius, by the way, is a figure who reminds us that, sometimes, even Nobel Prize winners make mistakes. Toward the end of the nineteenth century, he was one of the first to study the effects of CO_2 (carbon dioxide) on the global climate. In a famous article from 1896, he suggested the direct correlation between the concentration of CO_2 in the atmosphere and the temperature of our planet. He presented very detailed calculations and hypothesized that if the concentration of CO_2 were reduced by half, the average European temperature would fall by about five degrees Celsius, causing the continent to plunge into a new ice age. However, it was a risk he judged to be remote since, with the advent of the Industrial Revolution, the use of carbon as fuel was growing very rapidly and with it the concentration of CO_2. Indeed, in 1908, in his book *Worlds in the Making*, Arrhenius went so far as to highlight a positive aspect of all that burning of precious natural resources:

> We often hear lamentations that the coal stored up in the earth is wasted by the present generation without any thought of the future, and we are terrified by the awful destruction of life and property which has followed the volcanic eruptions of our days. We may find a kind of consolation in the consideration that here, as in every other case, there is good mixed with the evil. By the influence of the increasing percentage of carbonic acid in the atmosphere, we may hope to enjoy ages with more equable and better climates, especially as regards the colder regions of the earth, ages when the earth will bring forth much more abundant crops than at present, for the benefit of rapidly propagating mankind.

Let's say that there are unfortunately many other negative consequences that Arrhenius failed to foresee. . . .

Getting back to that December 10, Arrhenius opened his presentation of Einstein by recalling that "there is probably no physicist living today whose name has become so widely known as that of Albert Einstein. Most discussion [about him] centers on his theory of relativity." Then, however, he changed course and spoke of other things. Indeed, as strange as it might seem, Einstein won the 1921 Nobel Prize not for the theory of relativity but rather for another discovery undoubtedly less well known to the general public: the theoretical description of the photoelectric effect, another milestone in the story of quantum mechanics.

As with thermal radiation, in this case, too, our experience of daily life is a useful aid. Quantum mechanics sometimes shows up where you least expect it, even in the elevator. The photoelectric effect is commonly used in photocells, like the ones that keep elevator doors from closing when there is something, or someone, in the way. The effect happens when a metal surface is struck by an ultraviolet light and emits electrons, which detach from the material and can be measured and produce an electrical signal that, in this case, blocks the closing of the door. For there to be an emission of electrons, the light must be ultraviolet. With visible or infrared light, the photoelectric effect does not occur. This is impossible to explain with the classic wave theory of light.

To resolve the impasse, in 1905, Einstein, like Planck, set aside the tradition of classical physics and hypothesized that energy in the electromagnetic field was quantized. In addition, he introduced the concept of "quantum of light." In his paper of March 1905 in the *Annalen der Physik*, Einstein wrote: "The energy [of a beam of light] is not distributed continuously over ever-increasing spaces, but consists of a finite number of energy quanta that are localized in points in space, move without dividing, and can be absorbed or generated only as a whole." The quantum of energy is a photon, which resolves the contradictions between the photoelectric effect experiment and classical theory. Einstein assigned to the photon the energy *hf*, the same quantum that Planck had found for the thermal radiation of a black body. With this intuition, he formulated the theory that fully explained the photoelectric effect. Now the framework was finally complete. Not only is electromagnetic radiation produced in packets, as Planck had understood, but it can also propagate itself as a particle, that is, a photon.

"Owing to these studies by Einstein," Arrhenius concluded his speech, "the quantum theory has been perfected to a high degree and an extensive literature grew up in this field whereby the extraordinary value of this theory was proved."

Identity Crisis

Perhaps never in the history of physics has there been a cluster of new discoveries as abundant as the one in the decades around the turn of the nine-

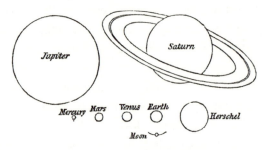

teenth century. Experiments that cast doubt on centuries of accumulated knowledge, new theories that revolutionized the description of the universe. The world of science was in continuous ferment, an ebullition of ideas that undoubtedly influenced the young Louis de Broglie, scion of the French nobility, who suddenly, after taking a degree in history, decided to make a 180-degree turn and devote himself to science, and specifically to physics. He began to practice the discipline during World War I, when he worked on the development of a system of radio communications for submarines. That was certainly not what made him end up in the history books, which he had so prematurely abandoned. Instead, it was his doctoral dissertation presented at the University of Paris in 1924.

De Broglie was fascinated by the recent findings of Einstein and Arthur Compton, which proved the corpuscular nature of light, and thus, its being simultaneously wave and particle. He hypothesized that the wave-particle duality could also be applied to matter. At that time, to associate wavelike properties with something as solid as matter was literally something out of science fiction. In effect, his thesis, though received with interest, was considered to be of little practical import. Yet no more than two years had gone by when, in 1926, two series of experiments confirmed de Broglie's hypothesis, for which he was awarded the Nobel Prize in 1929.

In line with quantum mechanics, de Broglie theorized a grand symmetry of nature. In sum, the universe is composed of matter and radiation, and both can behave either as waves or as particles.

Physics was now ready for the formalization of quantum mechanics, and in 1925, the Austrian physicist Erwin Schrödinger formulated the equation that is named after him and that describes the evolution of the quantum world. In the microscopic realm, the concreteness of objects is replaced by the uncertainty of probability.

The Apple and Mars

Two hundred and fifty years earlier, an apple fell from a tree in a garden in Lincolnshire. We do not know for a fact that it struck Isaac Newton, the celebrated English scientist who was said to have been meditating under the apple tree. It is certain that in those years, thanks in part to the previous studies of Galileo, Newton laid the groundwork for classical mechanics. This is the branch of physics that studies the equilibrium and the movement of bodies and that, until the end of the nineteenth century, was applied with great success in describing nature and the world, from astronomy to the machines of the Industrial Revolution.

Newton's mechanics allows us to determine the movement of an object once the forces that act upon it are known. This is done by way of the fundamental law

$$F = ma. \tag{2}$$

This law tells us that if we know the interaction of a body with its surrounding environment, that is, the force F, we can derive the acceleration a, which in essence means knowing its motion. Put simply, the elegance and power of this law lie in its statement that the motion of a body is completely and unequivocally determined by its relationship with the world.

In other words, the same force causes different motions depending on the mass of the body to which it is applied. This is something we all know quite well from experience. We need only think of the different effect we obtain by throwing with the force of our arms a soccer ball or a rock of the same size. Therefore, mass describes the inertia of a body with respect to the application of a force: the greater the mass, the less impact that a given force will have on the body.

Newton's law, as is true for all of classical mechanics, is deterministic: given a force and the characteristics of the body's motion in a specific instant, we are able to predict with absolute accuracy the trajectory of the body. This predictive capacity of the law is quite admirably displayed, for example, in the description of the motion of the planets or in space voyages. In July 1969, NASA scientists succeeded in taking two men to a precise point on the Moon after a voyage of more than 384,000 kilometers. In February 2021, the successors of those scientists guided the Perseverance rover onto the surface of Mars after a voyage of about seven months and 480 million kilometers, landing it precisely in the right place. All of this was thanks to mechanics, with which it was possible to calculate with extreme precision the trajectory of the spaceships.

Classical mechanics works, and it allows us to predict the future. But also not.

Crystal Balls and Smoke Signals

"Ibis et redibis non morieris in bello": "You will go, you will return, never in war will you perish," or "You will go, you will return never, in war you will perish." What a difference a comma makes! The Cumaean Sibyl was clever, but predicting the future has been a human ambition since well before the birth of science. Prophets, witch doctors, and fortune-tellers have always had an audience. The hope that, with enough acumen and the right equipment—whether it was animal innards, the smoke clouds of a campfire, or a crystal ball—one could foresee what had not yet happened has always been well rooted in humanity. It is not hard to imagine, therefore, how much this hope was nourished by Newton's *Principia*, published in 1687.

With Newton, the prediction of the future became science and no longer a wager or a question of interpretation. His equation of motion makes it possible to predict with certainty the position of a body. If to that we add the nineteenth-century development of electromagnetism, it, too, completely deterministic, and the fact that all systems are constituted by elementary building blocks, we can see how at the dawn of twentieth century the dream of predicting the future seemed to be at hand once we had acquired sufficient capability of calculation.

But they were jumping the gun. In the first decades of the twentieth century, physics, the very same discipline whose classical mechanics inspired the dream of deterministic predictability, began to undermine the very foundations of classical theory.

Not Only Cats

Erwin Schrödinger was a Nobel Prize–winning physicist who, in popular culture, is best known for his thought experiment about a cat. The fate of his hypothetical cat is ruled by a random subatomic event—a radioactive decay—that may or may not occur. Due to quantum superposition, the cat may be considered simultaneously both alive and dead. Despite the terrible reputation that he made for himself among cats all over the world, Schrödinger—who, by the way, apparently had a dog—made a fundamental contribution to the development of quantum mechanics with the equation that bears his name:

$$-\frac{\hbar^2}{2m}\nabla^2\Psi + V\Psi = i\hbar\frac{\partial\Psi}{\partial t}. \tag{3}$$

Don't let the apparent complexity scare you. Sure, a full understanding of it is something reserved for specialists in the field, but in reality this equation is in many ways analogous to Newton's equation, which we saw earlier. In this case, too, the starting point is the interaction of a particle with the outside world—here represented by potential energy V—which then serves as the basis for calculating the solution for predicting the future. Only in this case the object of the prediction is not the precise position of the particle but the probability of finding it in any given place. The function ψ, which we obtain by resolving Schrödinger's equation, describes, in fact, a wave in complete coherence with de Broglie's hypothesis. The wave function ψ does not tell us, however, precisely where the particle is, but only where probabilistically we could find it. The determinism of the classical mechanics of Newton's laws is ousted by the uncertainty of quantum mechanics.

This does not mean that classical mechanics is wrong; rather, it only works in certain realms. Quantum effects are visible only in the microscopic world. On the macroscopic scale—where macroscopic includes everything from a grain of sand to the planet—classical mechanics works just fine, as we have seen. Just as special relativity extends the area of validity of physical laws in conditions of high velocity, so quantum mechanics extends their validity in conditions of microscopic dimensions, that is, on the atomic or subatomic scale. And just as a universal physical constant—the speed of light—is the hallmark of relativity, so another physical constant, Planck's constant, is the signature of quantum effects. Note, for example, that it also appears in Schrödinger's equation (3).

Newton's law ($F = ma$) and Schrödinger's equation are thus basic tools with which physics describes the world, whether in the classical or the quantum version. Although separated by centuries and symbols of complementary worlds, they are also joined by what may look like a simple letter of the alphabet, m, which signifies a fundamental property of any object under study, whether it is a neutron or the Apollo 11 space capsule: mass.

In both classical and quantum physics, mass plays the crucial role of mediating the interaction of a body with forces, or rather with the world. In certain cases, other properties of a body also play this role. Electric charge and velocity, for example, mark the interaction with the electromagnetic field, but mass is present whichever the force might be.

The mass of bodies is also central in another fundamental physical process: universal gravitation. It was, of course, Newton who discovered how two objects, by reason of their having mass, attract each other through the force of universal gravitation, which is directly proportional to the masses of the bodies and gradually decreases as the bodies grow farther apart. Technically, the amplitude of the force of gravity between two bodies of masses m_1 and m_2 separated by distance r is expressed as

$$F_g = G\frac{m_1 m_2}{r^2}, \qquad (4)$$

where G is the constant of universal gravitation.

Gravity is one of the four fundamental forces. It is responsible for the motion of the planets and for the sensation of weight on Earth. Indeed, it is the attraction that the mass of the Earth exercises on any object in its vicinity and that determines the weight of the object. Weight, therefore, is nothing other than a force. Any object of mass m in the vicinity of the surface of the Earth is subject to a force of attraction P from the Earth calculated as

$$P = mg, \qquad (5)$$

where g is the acceleration of gravity, which depends on mass and the radius of the Earth and the gravitational constant G. This is also an expression of Newton's law $F = ma$, and again the mass of an object mediates its interaction with the world, in this case with our planet. Since the acceleration of gravity is nearly constant in proximity to Earth, in our daily experience weight becomes a synonym of mass.

That simple m, a fixed presence on shopping lists ("Remember to buy 3 kilos [6 1/2 pounds] of apples and 200 grams [1/2 pound] of cheese!"), in furtive glances at the bathroom scale after a lavish lunch, and on road signs, is actually a crucial ingredient of physics. Remember that the next time you're leaving to go on a trip and find yourself lugging an overweight suitcase filled with useless items (that's why they call it luggage). You are verifying a fundamental law of physics. That might provide you with some consolation.

October 21

On October 21, 1944, Max Planck was waiting anxiously, or perhaps with resignation, for the outcome of the trial of his son Erwin, which just two days later would conclude with the death sentence. On October 21,  1520, Ferdinand Magellan discovered the strait that bears his name. On the same date in 1879, Thomas Edison filed his application for a patent on the incandescent light bulb; in 1833, Alfred Nobel was born; in 1917, Dizzy Gillespie; and in 1995, Doja Cat.

A few minutes' research on the internet is enough to discover dozens of other events or famous birthdays that occurred on October 21. Considering that a year has 365 days and that history-making events are much more numerous, you don't have to be a statistician to realize that there is nothing special about October 21. Except with regard to the metric system: there are not millions of units but rather only seven, and yet the definitions of not one but two of these units were literally revolutionized on October 21. This is, to say the least, extraordinary.

As we have seen in the preceding chapter, on October 21, 1983, the 17th General Conference on Weights and Measures defined the meter in terms of the speed of light. Twenty-eight years later, on October 21, 2011, the 24th meeting of the same conference definitively decreed the end of an era: the oldest of the artifacts used for the definition of a fundamental unit of measurement, the proto-type kilogram, was retired. The solidity of a piece of precious metal—like all human works, inevitably transitory—was replaced by the solidity of nature, universal and available to everybody. This new solidity was obtained in part, in an apparent paradox, by way of the universal constant that had undermined the certainty of classical mechanics and that most evokes indeterminacy: Planck's constant.

Quanta and Balances

The new and revolutionary definition of the kilogram rests on two fundamental theories of modern physics: relativity and quantum mechanics. In both, as in general for all of physics, the concept of energy is central. In relativity, energy is expressed with what is probably the most popular formula in all of science:

$$E = mc^2. \tag{6}$$

This equation ties together energy E, mass m, and the speed of light c.

In quantum mechanics, energy enters quantum mechanics with the formula that expresses Planck's quantum, which we have been talking about in this chapter:

$$E = hf. \tag{7}$$

Pay attention! This is the same physical quantity, energy, that, thanks to relativity and quantum mechanics, can be expressed as a function either of the speed of light c, amply addressed in the preceding chapter, or of Planck's constant h. Energy thus becomes the bridge between relativity and quantum mechanics, and above all, it permits us to write mass as a function of two physical constants, which are universal and thus immutable, such as c and h. Far from being a mere artifice for specialists in the field, the relationship between mass and the two universal constants is of great practical use for the new definition of the kilogram. Indeed, when the need became evident to replace that perishable piece of metal that was the prototype kilogram with something more durable, physicists went to work and found various experimental methods for using the relationship that binds mass with h and c. Creating a unit of measurement necessarily involves the need to use it in practice. Specifically, if c and h are known with accuracy, it was necessary to think of an experiment that would permit the equally accurate measurement of mass and therefore to define the prototype kilogram.

The primary experimental instrument is the Kibble balance: a two-pan balance, no different in principle from those of 2000 BCE, just a little more technological. On one pan, you put the mass to be weighed, while the second balances the first. Rather than comparing the unknown mass to the weight of another mass, as in ordinary balances, the Kibble balance works by using an electromagnetic force. The value of this electromagnetic force can be measured with great accuracy by exploiting two quantum effects: the Josephson effect and the Hall quantum effect, and it is then expressed as a function of Planck's constant (omnipresent in quantum equations). If the value of the force is both fixed and known precisely—something that has been made possible in recent decades thanks to very refined experiments—the Kibble balance can measure the mass using the definition of the universal kilogram prototype, accurate and independent of material objects. h is truly small. Its value expressed in units of the international system is $6.626070150 \times 10^{-34}$, a number we can also write—if the publisher does not reprimand us for the consumption of ink—as 0.0000000000000000000000000000000006626070150. An international banking code number is child's play by comparison.

This is the measure that was chosen as the new definition of the prototype kilogram, which is based, therefore, on Planck's constant.

Small Pieces of Paper

On May 2, 1945, the Red Army raised the Soviet flag over the Reichstag in Berlin while Hitler committed suicide in his bunker. On May 8, Nazi Germany surrendered. The activities of the Manhattan Project, however, continued without pause. At 5:29 in the morning of July 16, 1945, in the desert of Jornada del Muerto, near

the city of Socorro, New Mexico, an artificial dawn lit up
the sky. It was the Trinity Test,
the first explosion of an atomic
bomb, the trial of ordnance that
a few days later would raze the
city of Hiroshima to the ground.
Among those present at the
Trinity Test was Enrico Fermi
(his typed eyewitness testimony,
which follows, is held at the Na-

tional Archives, RG 227, OSRD-S1 Committee, box 82, folder 6
"Trinity"):

> On the morning of the 16th of July, I was stationed at the Base
> Camp at Trinity in a position about ten miles from the site of the
> explosion.
>
> The explosion took place at about 5:30 A.M. I had my face
> protected by a large board in which a piece of dark welding glass
> had been inserted. My first impression of the explosion was the
> very intense flash of light, and a sensation of heat on the parts
> of my body that were exposed. Although I did not look directly
> towards the object, I had the impression that suddenly the coun-
> tryside became brighter than in full daylight. I subsequently
> looked in the direction of the explosion through the dark glass
> and could see something that looked like a conglomeration of
> flames that promptly started rising. After a few seconds, the rising
> flames lost their brightness and appeared as a huge pillar of smoke
> with an expanded head like a gigantic mushroom that rose rapidly
> beyond the clouds probably to a height of the order of 30,000 feet.
> After reaching its full height, the smoke stayed stationary for a
> while before the wind started dispersing it.
>
> About 40 seconds after the explosion the air blast reached me. I
> tried to estimate its strength by dropping from about six feet small
> pieces of paper before, during and after the passage of the blast

wave. Since at the time, there was no wind, I could observe very distinctly and actually measure the displacement of the pieces of paper that were in the process of falling while the blast was passing. The shift was about 2 1/2 meters, which, at the time, I estimated to correspond to the blast that would be produced by ten thousand tons of T.N.T.

A mass of 10,000 tons, 10 million kilograms of TNT. Fermi's estimate, done by dropping small pieces of paper, was not far off. The bomb released twice as much energy, equal to 22,000 tons of TNT. An enormous amount; just compare it to the energy released by one of the most powerful bombs dropped in World War II, the Grand Slam, which was less than the equivalent of 10 tons of TNT. Thanks to Einstein, nature had revealed to humanity that mass could be converted into energy. On that July 16, humanity had learned how to do it in a destructive way. Physics had lost its innocence. "Now I am become death, the destroyer of worlds," commented the director of the Manhattan Project, Robert Oppenheimer.

Fortunately, since then, reason has prevailed, and we have used the conversion of mass into energy expressed by the equation $E = mc^2$ only for peaceful purposes in fission power stations for the production of electricity. The future, furthermore, in a sort of positive counterpoise, could give us a new version of converting mass into energy that would contribute significantly to resolving the grave environmental crisis that we are now experiencing. Something that, as early as 1920, Arthur Eddington dreamed while studying the stars: "A star is drawing on some vast reservoir of energy by means unknown to us. This reservoir can scarcely be other than the subatomic energy which, it is known, exists abundantly in all matter; we sometimes dream that man will one day learn how to release it and use it for his service."

Today, we know that that process is nuclear fusion and that it actually does fuel the Sun and the stars, as Eddington hypothesized. In fusion, two lightweight nuclei of hydrogen, or its isotopes, combine, and in the reaction, part of the mass of the reactants is converted into energy. Scientists all over the world are now trying to steal the secret of the Sun in order to reproduce this process in the laboratory. It is a rough road, but much has been done, and the hope is that in a few decades we will have at our disposal a source of clean, unlimited electric power, free of CO_2 waste and ideal, therefore, for ensuring a sustainable future for our planet.

FOUR

The Kelvin

To Your Health!

"If we look at a glass of wine closely enough we see the entire universe." You might think such a statement was spoken after its author had already had an intimate look at several of his own full glasses. After all, it has been very well known since ancient times how much wine can favor inebriation and its consequent fantasies. Archaeological evidence of the first large-scale production of wine has been found near modern-day Tbilisi, Georgia, and dates back to 6,000 BCE. It might surprise you, therefore, that this apology for the glass of wine concludes a classic essay by the Nobel Prize winner Richard Feynman, entitled "The Relation of Physics to Other Sciences," published in his book *Six Easy Pieces*. In reality, however, the bond between wine and physics—and science in general—is much stronger than is generally thought. Feynman continues:

[In the glass of wine, there is] the twisting liquid which evaporates depending on the wind and weather, the reflections in the glass, and our imagination adds the atoms. The glass is a distillation of the earth's rocks, and, in its composition we see the secrets of the universe's age, and the evolution of stars. What strange array of chemicals are in the wine? How did they come to be? There are the ferments, the enzymes, the substrates, and the products. There in wine is found the great generalization: all life is fermentation. Nobody can discover the chemistry of wine without discovering, as did Louis Pasteur, the cause of much disease.

In sum, a glass of red wine, in addition to being a pleasure, is also a, still partially unexplored, scientific laboratory. Billions of sips and 8,000 grape harvests later, the detailed physical description of the process set in motion by rotating a glass of wine before bringing it to your mouth was the subject of an article published in 2020 in the prestigious journal *Physical Review Fluids* and subsequently excerpted in *Nature*. The scientist Andrea Bertozzi and her colleagues at UCLA studied "wine tears," those filmy arches that appear on the inner walls of a glass in which wine has been swirled.

Following the rotation of the wineglass, a thin film of liquid forms, from which droplets slither down the sides of the glass to then fall back into the wine. The phenomenon, well known to worshippers of Bacchus, occurs due to a combination of a property of the interface between two fluids and gravity. When the glass is put into rotation, wine deposits on its inside walls. The alcohol in this filmy layer evaporates more rapidly than that in the wine in the bottom of the glass, thus creating two fluids with different chemical properties: the original wine and the wine on the inner walls, which has a lower alcoholic proof, since part of the alcohol has evaporated. A physical effect takes place between these two fluids, which makes the fluid climb up the sides of the glass, where it accumulates until it falls back down with a configuration similar to tears. This process was described in 1865 by the physicist Carlo Marangoni from Pavia,

Italy, who elaborated on and completed an earlier study by James Thomson, older brother of the more famous William, about whom we will hear more later in this chapter. Marangoni's study left open the question as to why the wine slithers back down in a nonuniform way, forming little arches. One hundred and fifty years later, Bertozzi and her group came up with the explanation, thanks to a sophisticated theoretical model: the thin film of wine is formed by microscopic waves. It is these undulations that vary in thickness, together with gravity, that cause the tears.

The greater the alcohol content of the wine, the more visible the effect. So one might be induced to think that all a physicist sees in a bottle of Barolo or Amarone della Valpolicella is the opportunity for a new scientific article that further refines our understanding of those wine tears. All the more so since recent tendencies in research policy, summarized in the slogan "publish or perish," unfortunately promote the measurement of merit exclusively on the basis of the number of articles published rather than the quality of their content. Fortunately, that's not how it actually works, and it is masters like Feynman who help us keep our feet on the ground, or in this case our lips to the glass. His essay concludes, in fact, with these words: "If our small minds, for some convenience, divide this glass of wine, this universe, into parts—physics, biology, geology, astronomy, psychology, and so on—remember that nature does not know it! So let us put it all back together, not forgetting ultimately what it is for. Let it give us one more final pleasure: drink it and forget it all!"

From Sensations to Measurements

The advent of the scientific revolution, at the turn of the sixteenth century, brought two important innovations in the description of natural phenomena. The first was a shift toward abstraction and the replacement of qualitative descriptions with mathematical ones. This is summarized quite nicely by Galileo in *The Assayer* (*Il saggiatore*): "Philosophy [that is, natural philosophy] is written in this grand book—I mean the Universe—which stands continually open to our gaze, but it cannot be understood unless one first learns to comprehend the language and interpret the characters in which it is written. It is written in the language of mathematics, and its characters are triangles, circles, and other geometrical figures, without which it is humanly impossible to understand a single word of it; without these, we are left wandering aimlessly in a dark labyrinth."

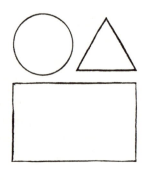

Galileo himself provides an admirable example of this revolutionary approach in his descriptions of motion and inertia, in which he examines motion apart from the complications of contingent effects, such as friction, concentrating on ideal properties.

The second innovation involved turning to measurement as an essential method to describe nature. This led to the development of new instruments, including those for the measurement of temperature, which today is one of the seven quantities of the international system and certainly one of the best known and most used. The sensations of the different gradations of hot and cold are intrinsic to human experience. Since ancient times, humanity has understood how influential temperature is in our lives, as well as in nature and its processes. Perhaps the most obvious example is the changing of the seasons. No wonder then that with the advent of the Renaissance, thermometry attracted the interest of scientists.

Galileo is credited with the invention, around 1592, of the first thermometer or, more properly, the thermoscope. This was a useful instrument for comparing the temperatures of two objects or for measuring variations in temperature, but it did not give absolute values. Galileo's thermoscope was essentially a glass tube open on one end and with a bulb on the other. Partially filled with water or wine, the tube was immersed, open end down, in a container full of the same liquid, so that the bulb end was filled with air. Putting the bulb in contact with the object whose temperature was to be measured caused the air in the bulb to contract or expand, depending on whether the temperature of the object was lower or higher than the ambient temperature. If the air contracted, the level of the liquid in the tube rose, indicating that the object was colder than the thermoscope. If, instead, the air expanded, bubbles were created, which gurgled in the liquid and caused its level to lower. This principle of measurement had been known since the time of ancient Greece, when Philo of Byzantium and Hero of Alexandria developed air thermoscopes. As in many other fields, however, the prevalence of Aristotelian theories put an end to the development of the sciences for more than a millennium.

Nevertheless, by Galileo's time, there was a need for different instruments to give the same reading when used in the same circumstances. In the case of thermometry, one solution was to use instruments that were exactly the same, though at the time this was not simple to do. Another possibility, much easier, was to make the readings of different instruments comparable, using a common point of reference. The basis is the principle of causality, according to which similar effects have similar causes. In the case of temperature, one starts with the observation that a certain thermometer always gives the same reading every time it is put into contact with different samples of melting ice, which is the point of reference in this case. From the constancy of the effect (or the consistency of the thermometer reading) one can deduce the constancy of the cause. This leads to the conclusion that the same phenomenon, characterized by a constant temperature, is at work in the various samples of melting ice. Consequently, if another thermometer is immersed in melting ice, it will produce the same temperature reading that was recorded on the first thermometer in a similar situation, because the same cause must always have the same effect.

Among the first to use a scale based on points of reference was the Venetian Giovanni Francesco Sagredo, a close friend of Galileo. Sagredo built a series of air thermometers, which he claimed produced identical results. Using them, he provided quantitative measurements, establishing a scale that read 360 degrees at the apex of summer heat, 100 degrees in snow, and zero degrees in a mixture of snow and salt. It was known that salt water freezes at temperatures below that at which pure water does, 0°C (32°F). It seems reasonable, therefore, that Sagredo chose snow and the mixture of snow and salt as points of reference for fixing his temperature values. Sagredo showed his passion and enthusiasm for the new horizons opened up by the measurement of temperature in a letter he wrote to Galileo in 1613: "The instrument for measuring heat, invented by Your Excellency, has been reduced by me in various convenient and exquisite forms, so that the difference of the temperature from one room to another is seen up to 100 degrees. I have with these speculated about a number of wondrous things, as, for example, that in winter the air is colder than ice or snow."

We owe the application of the thermometer to measuring body temperature to Santorio Santorio, born in 1561 in Capodistria (modern-day Koper, Slovenia), then a dominion of the Most Serene Republic of Venice. An acquaintance of Galileo, he was called to teach medicine at the University of Padua in 1611, where up until one year earlier Galileo himself had also been on the faculty. Santorio was one of the pioneers of the use of quantitative physical measurements in medicine, bringing to this discipline the experimental method with which, in those very same years, Galileo was revolutionizing science. Inspired by Galileo's findings on the motion of the pendulum, Santorio invented the pulsilogium, a device for measuring the heartbeat. He was also the first physician to observe variations in human body temperature and to interpret them as indicators of health or illness. Santorio modified the air thermometer, inserting the bulb into the mouth of the patient. For the gradations on the tube, he used two points of reference: the temperatures of snow and the flame of a candle.

A Question of Equilibrium

The common thermometer is an excellent example of the physical principles underlying the measurement of temperature. Let us start with principle of thermal equilibrium.

The concept of equilibrium is crucial for physics. In general, a system is in equilibrium when the quantities that characterize it do not change over time. The most familiar form of equilibrium is static mechanical equilibrium, which we achieve, for example, by laying our cell phone on the table. The telephone remains immobile—that is, its position does not vary over time—because there are no forces that make it move or rotate. Mechanical equilibrium, however, is not the only kind. The battery of the switched-on cell phone sitting on the table is in mechanical equilibrium, but it is not in chemical equilibrium. Reactions are going on inside it, which produce electrical energy and change the phone's condition (its battery loses its charge). Then there is a third kind of equilibrium, thermal, which occurs when the temperature of an object remains constant. Temperature is the physical quantity that characterizes thermal equilibrium. If two systems at different temperatures enter into contact, they will reach thermal equilibrium; in other words, their temperatures will equalize as the warmer body transfers heat to the cooler one.

The thermometer is an instrument of measurement that enters into thermal contact with another system and places itself in thermal equilibrium with it by assuming the same temperature without modifying it. For this to happen, the contact point of the thermometer must be narrow, so as not to significantly disrupt the system being measured. This is how a body-temperature thermometer works: it rapidly assumes thermal equilibrium with the human body without altering the body's temperature. The thermometer makes this temperature variation visible thanks to a thermometric characteristic, or in other words, a physical process that varies as a function of temperature and that can therefore be easily observed. In traditional thermometers, this process is the thermal expansion of the liquid inside it (once upon a time it was mercury, today it is alcohol or an alloy of gallium), and the thermal characteristic is the liquid's length. This works because for most substances a body's volume tends to grow as the temperature increases, and consequently its dimensions vary. This is a phenomenon well known to those who design bridges, buildings, or railroads and is compensated for by special dilating joints. The length of the column of liquid in the thermometer varies in relation to the temperature of the body with which it is in contact; thus, the measurement of temperature is transformed into a measurement of length, much easier to visualize. In electronic body-temperature thermometers what varies with the temperature is instead an electric resistance whose value is digitalized and displayed on the thermometer's little screen.

Boiling Water and Melting Ice

In the same period that Santorio was teaching humanity how to take its temperature, Swedish engineers were at work building what was believed to be one of the most powerful warships of the time, the *Vasa*. On August 10, 1628, an amazed crowd lined the docks of the port of Stockholm where, in the presence of the king, the ship was launched. But the crowd's enthusiasm soon turned to dismay. Less than a mile from the launch ramp the *Vasa* suddenly sank, owing to what were supposedly harmless gusts of wind, taking 30 crew members down with it. The ship was armed with 64 bronze cannons, distributed on two decks. The upper deck of cannons, added at the wish of the king, made the ship too tall with respect to its width and therefore very unstable. Another flaw was that the wood structure of the *Vasa* was thicker on its left side than on its right. The ship's carpenters appear to have used different systems of measurement. Indeed, archaeologists have found four rulers used by the workers who built the ship: two are calibrated in Swedish feet, which had 12 inches, while the other two  are calibrated in Amsterdam feet, which had only 11.

The *Vasa* fiasco is just one of many cases in which the use of different scales of measurement for the same project has led to miserable failures. We have already mentioned the Mars Climate Orbiter, which disintegrated in the atmosphere of Mars because some engineers had used metric units while others had used English units, but similar stories abound. One example is Air Canada flight 143, which was scheduled to fly on July 23, 1983, between Montreal and Edmonton with an intermediate stop in Ottawa. Due to a series of unfortunate circumstances, including the failure of the automatic fuel quantity indication system, the plane was refueled with fuel that had to be calculated by hand using a dipstick inserted in the tanks. The dipstick reading was in centimeters, which had to be converted into liters and finally into kilograms, using the appropriate conversion factors. But in the final calculation, the conversion was made from liters to pounds instead of kilograms. Because of this error, along with other issues, the plane could not reach its destination and

was forced to make an emergency landing—luckily without consequences for the passengers and crew—on a motor racing track
on a former Royal Canadian Air Force base in Gimli, Manitoba
(the location has since given this flight the popular name of "Gimli
glider"). Reaching agreement on units of measurement is no simple
matter, as we have seen in the preceding chapters. Temperature is
no exception: witness the division in the world today between the
use of Fahrenheit and Celsius.

Less than a century had gone by since the development of the
first thermoscopes when scientists began to conceive of a universal
scale for the measurement of temperature. Newton and the Danish
astronomer Ole Rømer were among the first to come up with universal scales in the early 1700s, but it was not until 1724 that Daniel
Fahrenheit proposed the scale that bears his name, still used today
in the United States. He also gets the credit for the idea of using
mercury as a thermometric liquid, a transformative choice. Thanks
to its elevated coefficient of expansion, the use of mercury significantly improved the accuracy of thermometers. For a given temperature variation, mercury expands much more than water or alcohol,
thus allowing for a more accurate visualization of the temperature
itself. Fahrenheit chose as reference points the temperature of a
solution of water, ice, and ammonium chloride, to which he assigned
a value of 0, and the average temperature of the human body, which
he fixed at 96, also noting, however, the value of 32 for the temperature of melting ice. Today, the Fahrenheit scale, the official scale
in the United States, is based on two reference points separated by
180 degrees (indicated with °F): the temperature of melting ice,
fixed at 32°F, and that of the boiling point of water, 212°F.

In 1742, the scientific world witnessed the debut of an alternative scale, which would come to be dominant, this one invented by
the Swedish astronomer Anders Celsius. In an article that would
later become famous, Celsius defended the two reference points
he chose for his thermometric scale (originally selected by Santorio

but not yet universally accepted). Celsius identified the points in the temperature of melting ice and boiling water in standard conditions of pressure, separating them by 100 degrees. Originally, Celsius chose to assign the value of 100 to melting ice, and 0 to boiling water, but the convention was reversed shortly after his death. Today, this system bears his name and the indication °C.

A Six-Pack of Ice-Cold Peroni

In the late 1970s, Italy fell in love with Fantozzi, the endearingly hapless white-collar time-card puncher created by comedian Paolo Villaggio, and the hero of nine extraordinarily successful movies. In the second film in the series, the off-screen narrator introduces a typical Fantozzi moment, as the beleaguered bookkeeper returns home from the office and prepares to watch the broadcast of the Italy versus England soccer match. "Fantozzi had a fantastic plan: socks, underwear, flannel bathrobe, tray table in front of the television screen, mouth-watering super-thick onion omelet, a six-pack of ice-cold Peroni, berserk cheering, and unbridled burping."

In Italy, if you want to talk about the temperature for serving beer, a tribute to the great Paolo Villaggio and ice-cold Peroni is obligatory. It is certainly true that in the 1970s, beer by definition was meant to be drunk cold. (Cue the inevitable comments after a trip to England about the lukewarm beer in the pubs.) Since then, however, the enormous evolution of the range of beer offerings and the consequent refinement of beer drinkers' palates have accustomed us to serving temperatures that range from around 0°C to nearly 16–18°C, depending on the brew. A rich literature has grown up around this theme, which has received attention even from a high-brow newspaper like the *Wall Street Journal*. With all due respect to Fantozzi and ice-cold Peroni, the thermometer has become an indispensable instrument for the proper enjoyment of a glass of beer. Naturally, in this case, too, units of measurement must be used with caution, given that in Europe a Pilsner is supposed to be imbibed at around 4–6°C, while in the United States the same beer is served at 38–45°F. In case of error, there are obvious repercussions for the taste experience.

The relationship between temperature and beer actually goes much deeper, proof that wine is not the only beverage with an important stake in this particular physical quantity. Temperature is a fundamental quantity in thermodynamics, the branch of physics that studies macroscopic processes involving exchanges of energy between systems and their environment, the transformation of mechanical work to heat, and, vice versa, of heat into mechanical work. Heat is a form of energy. More precisely, it is the energy transferred between two bodies at different temperatures, with the warmer body transferring energy to the cooler and thus becoming cooler itself.

One of the fathers of this discipline is James Prescott Joule, an English brewer from Salford, in Lancashire. Joule is credited with the demonstration, in the 1840s, of the equivalence between mechanical work and heat, both of which are mechanisms for transferring energy to a system. In one famous experiment, Joule showed that the temperature of water in a container can be raised by using a mechanical process, specifically, by making a sort of propeller rotate inside it. The mechanical energy used to keep the propeller turning is converted, thanks to friction, into thermal energy in the water. Joule laid the groundwork for modern thermodynamics and, in particular, for the basic principle of the conservation of energy—the first law of thermodynamics—disproving caloric theory. Caloric was thought to be a kind of invisible and immaterial self-repellent fluid that could flow from hotter to colder bodies and whose concentration explained an object's higher or lower temperature. Instead, Joule demonstrated that heat, too, is a form of energy transfer. He obtained his results thanks to highly accurate measurements of temperature, the fruit of his supreme experimental skill, which is said to have derived from his practice of brewing and, therefore, from his familiarity with chemistry and instrumentation. In the cemetery of the suburb of Brookland, south of Manchester, his tombstone is engraved with the number 772.55, which corresponds to his most accurate measurement—performed in 1878—of the factor of equivalence between mechanical energy and heat (expressed, naturally, in English units, the foot-pound force per British thermal units . . .).

Beer Molecules

A mug of beer contains about 10^{25} molecules. The superscript in 10^{25} is how physicists and mathematicians abbreviate a number composed of a one followed by twenty-five zeroes. Ten million billion billion molecules in a few gulps. When we taste the first gulp and our brain ponders whether the temperature is right or not, we hardly think that the sensation is correlated with that dreadful number of molecules. And we certainly do not consider that if the beer is too warm with respect to our expectations, this happens because those molecules are moving faster than the optimum speed for our palate. But it's true; the macroscopic thermodynamic properties of a physical system, such as temperature, pressure, and volume, are strictly bound up with the microscopic properties of its components.

The bond between macroscopic thermodynamic quantities and the microscopic behavior of matter is described by kinetic theory, of which Joule was one of the pioneers. This theory was perfected around 1870 by the Viennese physicist Ludwig Boltzmann. The prototypical system for kinetic studies is a gas, described as a set of microscopic particles (atoms or molecules) in constant and rapid movement inside a container. According to kinetic theory, the pressure, volume, and temperature of a gas are determined by the characteristic motion of the atoms or molecules of which it is composed. Pressure is caused by the random collisions of the particles against the container walls, and the temperature of the gas is tied to the kinetic energy (the energy of movement) that they possess.

Boltzmann's name has been given to the universal physical constant k_B, which binds the macroscopic world to the microscopic one and which appears in the equation that ties the kinetic energy of the particles of an ideal gas to its temperature:

$$E = 3/2 \; k_B \, T. \tag{1}$$

This formula states that the temperature T of the room in which you are reading these lines is proportionate to the average kinetic energy of the molecules of air E, or the square of the average velocity of those same molecules. The hotter the air, the faster the molecules move. The constant of proportionality between energy and temperature is k_B, which has a universal value equal to 1.380649×10^{-23} in the international system of units. At room temperature, the air molecules move at about 1,800 kilometers per hour!

The extraordinary elegance of physics lies in part in its ability to express the bond between the infinitely small and the macroscopic world, between an atom and a blimp, by way of a simple formula of just a few characters, as in equation (1). This is crucial to expressing temperature with an appropriate scale, a scale that bears the name of a small waterway and that was proposed in 1848.

Brothers

The history of physics is filled with strange anecdotes. Like Joule, Boltzmann also had a physics formula engraved on his tombstone, namely, the equation for entropy. And if Harald Bohr's notoriety as a soccer player didn't manage to overshadow that of his brother, Niels, the same applies to the brothers Thomson. James and William were born two years apart in Belfast, James in 1822 and William in 1824. James was a scientist and inventor. He was the one who initiated the study of wine tears that we discussed in the opening paragraphs of this chapter. Even the most expert enologists probably do not remember him, however, since the description of those alcohol tears was attributed to the Italian Carlo Marangoni, who perfected it. As in the case of the Bohrs, only one of the Thomson brothers entered the pantheon of physics, and it was not James. For his scientific merits, the younger brother was the first English scientist to be named to the House of Lords, receiving the title of Baron Kelvin. The title refers to a little stream 35 kilometers long that flows north of Glasgow and whose worldwide fame is due to its flowing by the laboratory of William Thomson.

Thomson was an eclectic scientist and was involved in laying the first transatlantic undersea telegraphic cable. His fame, however, is primarily tied to thermodynamics, and he is credited with the introduction, in 1848, of the temperature scale that bears his name. Although, in terms of everyday use, it is not as well known as the Celsius and Fahrenheit scales, the Kelvin scale is a keystone of thermodynamics since its definition is independent of the properties of a substance, such as water or the human body. The unitary increment of the Kelvin scale is identical to that of the Celsius scale, but instead of fixing zero to the temperature of melting ice, it defines zero as the coldest possible point for matter (-273.15°C). Nothing can be colder than this "absolute zero." The Kelvin scale, therefore, is an absolute scale and describes the amount of movement energy of the microscopic components of matter: atoms and molecules. In that sense, temperature expressed in kelvin is exactly that which is used in the formula $E = 3/2 \, k_B \, T$, mentioned in the previous section.

The kelvin (indicated as K) has been the base unit for thermodynamic temperature since 1954, when the General Conference on Weights and Measures adopted it. To make a unit of measurement concrete, or rather to convert its definition into practice, you need an experimental method. For the unit of temperature, the procedure sets out not to achieve 1 K but rather to attain 273.16 K, fixed at the triple point of water. This is when water coexists in thermal equilibrium in its three phases: solid, liquid, and gas. This is a valid universal standard because, at a given pressure, the triple point always occurs at exactly the same temperature, which is precisely 273.16 K. Until 2019, the kelvin was defined, therefore, as "the fraction 1/273.16 of the thermodynamic temperature of the triple point of water."

As with the other units of measurement, everything changed with the revision of the international system in line with universal physical constants. In 2019, the kelvin was redefined using Boltzmann's constant, now determined with extreme accuracy. This definition is based on the assumption that the fixed numerical value of Boltzmann's k_B constant is 1.380649×10^{-23} kg m^2 s^{-2} K^{-1}, where kilogram, meter, and second are identified in terms of the fundamental constants that we saw earlier. However, though it no longer defines the kelvin, the triple point of water still remains a convenient and practical way to calibrate thermometers.

The kelvin is the unit of measurement of temperature universally used in physics, but Celsius is the dominant scale in everyday life and in multiple practical applications. Whether because of tradition, the elegance and simplicity of its two easily remembered reference points—0 for freezing water (melting ice) and 100 for boiling water—or because it expresses many of the temperatures used in daily life in the small numbers we tend to prefer, the Celsius scale is used practically everywhere in the world today. Only the United States, some Pacific islands, the Cayman Islands, and Liberia use Fahrenheit as their official temperature scale.

An Unreachable Goal

The record for the lowest temperature ever recorded in the contiguous United States is -57°C (-70°F), recorded on January 20, 1954, at Rogers Pass, Montana, which cuts through the Rocky Mountains. If that seems cold, it's nothing compared to the world record, which appears to have been set in the Antarctic at the Russian Vostok research station on July 21, 1983, with a recorded

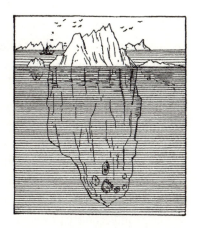

temperature of -89.2°C (-128.6°F). But even that seems rather mild compared to the -240°C (-400°F) measured by NASA's Lunar Reconnaissance Orbiter space probe in a crater near the lunar South Pole. Yet -240°C is still a long way from absolute zero. Physics laboratories, however, are able to obtain temperatures very close to it. In 2014, for example, at the Gran Sasso laboratories of the Italian National Institute for Nuclear Physics, researchers recorded a temperature of six millikelvins, a remarkable result because it was obtained in a relatively large volume of one cubic meter. In much smaller volumes, even lower temperatures are reached, just a few hundred billionths of a kelvin above absolute zero.

Temperatures close to absolute zero are of interest to physicists because, in those conditions, the behavior of matter is very different. At those temperatures, the thermal, electric, and magnetic properties of many substances undergo remarkable changes. Two important phenomena that occur under certain critical temperatures are superconductivity and superfluidity. Superconductive materials do not pose resistance to the flow of electricity and are thus used, for example, when it is necessary to generate intense magnetic fields, as in the LHC (Large Hadron Collider) particle accelerator at the CERN (European Organization for Nuclear Research). The LHC, in fact, has over 1,700 magnets that keep the particles on the right trajectories, all made of superconductive material and some of which weigh as much as 28 tons.

Regardless, absolute zero is a theoretically unreachable goal. The third law of thermodynamics states that as the temperature approaches absolute zero, it becomes more and more difficult to remove heat from a body, and thereby cool it. Reaching absolute zero in a finite time and using finite energy is therefore impossible. Quantum mechanics, emerging with the concept of probability out of the certainty of classical mechanics, also presented an issue. Heisenberg's uncertainty principle states that an experiment, no matter how accurate it may be, can never exactly determine both the position and the velocity of a particle, or to be more precise, its momentum, as given by the product of velocity and mass. This principle also dictates a limit to the accuracy with which the energy of a system can be determined during a certain time of observation.

In other words, the product of the precision with which you can measure the energy in a system ΔE and the duration of the interval of time Δt during which the measurement is performed cannot go below a certain limit. This is described in the formula $\Delta E \cdot \Delta t \geq h/4\pi$, where h is Planck's constant, which we have discussed in previous chapters. Establishing that a system is at the temperature of absolute zero would involve, therefore, determining with absolute precision that its energy is zero ($\Delta E = 0$), something that could be done only by hypothesizing an unrealistic infinite time of observation. From another point of view, taking an object to absolute zero would mean precisely stopping each of its atoms in a distinct point. This would require fixing the exact position and the exact quantity of motion of those atoms, which is again a contradiction of quantum mechanics.

Hotter Than the Sun

On March 2, the supersonic Concorde jet made its maiden voyage. At its cruising altitude, some 17,000 meters (55,000 feet) above sea level, the temperature was about 57°C (70.6°F) below zero. On July 20, humankind took its first steps on the lunar surface. During Neil Armstrong and Buzz Aldrin's sojourn the temperature oscillated between −23°C and 7°C (−9.4°F and 44.6°F). On August 15, the mud of Woodstock welcomed a generation of young people with the music of Joan Baez, Janis Joplin, and many others. The daytime temperature on those days was around 28°C (82.4°F), and it dropped all the way to 12°C (53.6°F) at night. It appears, however, that few of those present noticed. We are speaking, naturally, about 1969. In the spring of that year, a handful of English scientists left for Moscow, carrying in their luggage a thermometer to measure a temperature of 10 million degrees. A thermometer that demonstrated how science could be an instrument of peace, even at the height of the Cold War.

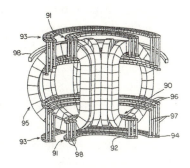

In those years, the tension between the two blocs, Western and Soviet, was red hot, and the nuclear arms race knew no rest. In just nine years, between 1960 and 1969, the USSR and the USA had conducted 660 bomb tests, and the world was living in a state of terror. Parallel to this military research, however, progressed a research program for the peaceful exploitation of nuclear energy. The first process to be studied and put into practice was nuclear fission, in which a heavy nucleus, struck by a neutron, divides, and in so doing liberates energy. In 1951, EBR-1, the first experimental fission reactor able to produce electric energy, began operation in the United States. The energy generated by EBR-1 could illuminate just four 200-watt light bulbs, but it was a historic event. In the Soviet Union, on June 27, 1954, the first civil nuclear power station was activated with the evocative name of Atom Mirny (peaceful atom). One year later, in Arco, Idaho, the reactor BORAX-III was already able to light up an entire city. In Europe the Calder Hall power station near Seascale, England, began operation in 1956, while the first fission power station in Italy entered into service in 1963.

Shortly after the end of World War II, however, an alternative process to fission came under serious study: nuclear fusion. As we have seen in our discussion of the kilogram, in the fusion process two lightweight nuclei of hydrogen isotopes combine, thanks to the prevalence of nuclear interaction. In the reaction, part of the mass of the reactants is converted into energy. As in the case of fission, the energy liberated by a single reaction is much greater than the energy obtainable from normal chemical combustion reactions, and no CO_2 is produced, but the enormous advantage of fusion is that there is no production of long-lasting nuclear waste. The process is intrinsically safe, and the fuel (water and lithium minerals) is substantially unlimited. It is no surprise that research of this kind takes on strategic importance. Therefore, at the height of the Cold War, the competition between the two blocs with regard to fusion was extremely, even dangerously, intense. Being able to exploit such an energy source would amount to an immense economic and political advantage. In light of this, when, in the summer of 1968, on the occasion of the Third International Conference on Plasma Physics and Controlled Research on Nuclear Fusion, the Soviets announced that the temperatures of the fuel in their experiment had reached 10 million kelvin, a lot of people in the West broke out into a cold sweat.

The reason for all the agitation was precisely the physics of fusion. To make the process happen, it is necessary to heat the two hydrogen isotope nuclei to very high temperatures in order to overcome the natural repulsive force that exists between them. Both nuclei, in fact, have a positive charge, and because of this electrostatic force—the Coulomb force—they tend to repel each other. If, however, they manage to get sufficiently close, they combine due to attractive nuclear forces. To get them close enough to each other, they must be heated to millions of kelvin so as to exploit the motion of thermal agitation, which we discussed earlier. At elevated temperatures, the nuclei move very fast, and therefore they have sufficient kinetic energy to overcome the repulsive barrier. At those temperatures, matter reaches the so-called status of plasma, an ionized gas, which is, in fact, the fuel of fusion. Scientists aim to heat the plasma of future fusion reactors to a temperature of 150 million degrees, a temperature greater than that at the center of the Sun.

For plasma, you need to have a container able to bear elevated thermal charges, whose walls do not degrade. To meet this need, the physicists who work on nuclear fusion have designed special doughnut-shaped steel containers in which the plasma is confined by an intense magnetic field. The magnetic field exercises a force on the charged particles in motion that compose the plasma.

One of the main experiments of the 1960s was called T-3, conducted at the Kurchatov Institute of Atomic Energy in Moscow. To perform the experiment, the Soviet researchers created a special configuration of the magnetic field using a machine called a tokamak, conceived a few years earlier by their compatriots Andrei Sakharov and Igor Tamm. In 1968, after several years of experiments, the scientists declared that they had heated the plasma of T-3 to as much as 10 million kelvin. Taking into consideration also other parameters of plasma, this was a spectacular result, which would have given the Soviets temporary supremacy in such a strategic field. Fortunately, despite the political divisions of the period, the scientific channels between East and West had remained open. So it was that, in accordance with the best scientific spirit, an independent evaluation was proposed. The Soviet physicists, aware of the value of their results, invited their British colleagues at the Culham Laboratory to come to Moscow in person to measure the temperature of their experiment. In addition to representing the "competition," the British possessed a very special thermometer, since they were experts in the measurement of the temperature of plasmas by using the laser, then only recently invented.

Coming at the apex of the Cold War, the Russians' proposal was bold and anything but simple. The political and diplomatic implications and difficulties were daunting, but both sides expected great benefits from the enterprise. For the Soviets, it would provide confirmation of their measurements and their supremacy. For the British, in contrast, it was a spectacular testing ground and an international stage for their applied physics and, in particular, for the technique of temperature measurement known as Thomson scattering, which they were perfecting in those years. This was a difficult technique based on tracking the light of a laser beam scattered by the electrons in motion inside the plasma.

Despite the mutual diffidence and the complications, the mission came off. The group of British scientists departed for Moscow, accompanied by five tons of equipment. After weeks of preparation, their measurements were successful and confirmed what their Soviet colleagues had reported the previous year, opening the way to the international success of the tokamak configuration. Just a few months later, the United States transformed its main experiment in the Princeton laboratory into a tokamak and quickly obtained similar results. In short, the tokamak configuration became the leading player in worldwide research on controlled thermonuclear fusion.

And science demonstrated that it could knock down walls.

FIVE

The Ampere

Scratch and . . . Volta

 When he began to scratch away with a fingernail the last three letters of the inscription adorned by a laurel wreath, more than a few were shocked. After all, they were at the library of the Academy of Sciences of the Institut de France, in Paris, and that inscription was dedicated to the famous Voltaire. It seems, however, that no one dared stop First Consul Napoleon Bonaparte. After all, his was not an act of gratuitous vandalism but a recognition of the Italian physicist Alessandro Volta, for whom the future emperor nurtured a great admiration. Thanks to his scratching, the inscription "To the great Voltaire" became, in fact, "To the great Volta." The veracity of this anecdote, recounted by Victor Hugo in his *William Shakespeare*, is debatable, given that no other direct testimony has come down to us. What is certain, however, is the high esteem in which Volta was

held by Bonaparte, who awarded him a medal of the Academy of Science, appointed him senator of the newborn Kingdom of Italy in 1809, and conferred on him the title of count.

This well-deserved esteem came in recognition of an excellent scientific career and, above all, for the invention of the electric battery. Born in Como in 1745, Volta played a pioneering role in the study of electrical phenomena (whose systematic investigation by the scientific community was only beginning at the time) and in the discovery of methane, which he identified in the swamps around Lake Maggiore. In 1799, he created the first electric battery, the voltaic pile, which, using chemical reactions, converted chemical energy into electricity. To get an idea of the enormous impact of Volta's discovery, just think of the quantity of batteries we use in everyday life and their increasingly important role in the electric economy of the future. That impact was recognized by Einstein himself, who, in 1927, on the occasion of the hundredth anniversary of Volta's death, dubbed the battery "the fundamental basis of all modern inventions."

The Missing Letter

At Corso Re Umberto 60, Turin, a plaque marks the birthplace of the person who "simplified the everyday life of writing." That person is Marcel Bich, born in the Piedmontese capital in 1914 before moving with his family to France. There, after World War II, he acquired and perfected the patent of the Hungarian inventor Bíró and started production of what would become probably the most common and widely used instrument of writing in the world: the Bic pen.

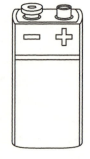

László József Bíró was the mastermind of the ballpoint pen, which, in Italian, bears his name and which freed writers from the fountain pen, with its need for cumbersome refilling operations. Bíró's pen soon found a buyer in the British Royal Air Force. Ink pens, in fact, were ill-suited to flying because they leaked too easily, while the new ballpoint pens—renamed Eterpen by the RAF—were ideal for quick note-taking by pilots. The Hungarian inventor, however, was unable to reach broader markets, whereas Bich did so with great success, thanks in part to such improvements as the transparent body that enabled users to check the amount of ink remaining anytime they wished.

Bich and Volta both became household names, but names which were only rarely associated with the actual people or their stories. The authoritative British daily the *Guardian* has estimated that, since the 1950s, something like 100 billion Bic pens have been produced, which would be enough to form a line to the Moon and back 320,000 times. It is no exaggeration to say that every inhabitant of the planet has held a Bic pen in their hands at least once in their lives and that only a small fraction of them have ever thought of that missing "h" or the pen's Piedmontese inventor. The International Energy Agency (IEA) estimates, on the other hand, that some 90 percent of the world's population has access to electricity, which means that at least 7 billion people have heard of the volt, the unit of measurement of the electric potential difference, often also defined as voltage.

It is precisely the electric potential difference between two points of a circuit that allows electric charges—in practical applications, electrons—to move along the circuit itself and to generate electric current: electric current that typically activates appliances connected to the circuit, whether they are light bulbs, radios, personal computers, cell phones, or citrus squeezers. Transporting electricity along the North American main lines takes up to 500,000 volts. Elevated potential differences, in fact, ensure more efficient transport, but for domestic use, transformers reduce the voltage to 220 volts, a more manageable figure commonly used in most of the world except in North America, where 110 volts is used. An ordinary AA battery delivers a potential difference of 1.5 volts, while a car battery delivers 12. In sum, for better or worse, the volt is a unit of measurement with which everyone, sooner or later, comes into contact. But, as in the case of Bich, it is likely that most of the world's users of electricity are not aware of the removal of a letter from Volta's surname (except perhaps in Italy).

Whether aware or not of the origins of the volt, the fact remains that when we talk about electricity in our daily lives, the volt is probably the best-known unit of measurement along with the kilowatt-hour. We see the volt particularly in safety warnings and the kilowatt-hour or megawatt-hour in more economic contexts. The kilowatt-hour, in fact, is the unit of measurement used to quantify the electricity charged on our monthly bill. In common parlance, however, the volt is used with a frequency similar to that of the kilogram, the meter, and the second. But while the public's knowledge of these latter three comes mainly from their scientific roles, since they are also fundamental units of measurement in the international system, that is not true for the volt.

The Eiffel Tower

While the celebrity attained by Volta thanks to the scraping away of the last letters of Voltaire is only an anecdote, that of André-Marie Ampère is much more solid, his surname being engraved in perpetuity—along with those of 71 other illustrious French scientists—under the second-floor balcony of the Eiffel Tower. A well-deserved honor,  given that Ampère, born in Lyon in 1775, was one of the trailblazers of electromagnetism. Gifted in both experimental physics and mathematics, Ampère was responsible for fundamental contributions to our understanding of electromagnetic fields. His mathematical skills were also displayed in an early study entitled *Considerations on the Mathematical Theory of Games*, in which he demonstrated that, in a game based on probabilities, a player was inevitably bound to lose against the bank.

In addition to the permanent celebrity of his name being etched in the iron of the Eiffel Tower, Ampère's discoveries earned him the honor of having a unit of measurement named after him: the ampere (with a small "a" and no accent), which has been included among the seven fundamental units of the international system. Electromagnetic phenomena are tied to electric charges and currents. On a microscopic level, matter has a property, its electric charge, which is rarely appreciable in macroscopic dimensions. Consider an atom. It is made up of a nucleus, which is, in turn, composed of particles called neutrons, protons, and electrons, which in Bohr's semiclassical model, orbit around the nucleus (we talked about this in the chapter on the meter). Protons and electrons differ in mass (a proton is about 1,836 times heavier than an electron), but also with regard to another property, electric charge. Protons have a positive charge, while electrons have an identical but negative charge. Neutrons have no electric charge. This is the basis of many fundamental properties of matter: particles with same-sign charges repel each other, while particles with opposite-sign charges attract each other. At macroscopic dimensions, all of this is hardly visible because single atoms have an equal amount of positive and negative charge, and bodies, therefore, are typically neutral.

When electric charges move, they give rise to currents. When we put a battery into a flashlight, the electrons are set in motion by the potential difference (the 1.5 volts from before) and circulate in the copper wires and bulb of the flashlight. It is the electric current that lights up the light bulb, just as it is the electric current flowing in the wires connecting the refrigerator to the outlet that makes the refrigerator work. Electric charges and currents are crucial for physics and for technological applications because they are also sources of the omnipresent electric and magnetic fields. The fundamental unit of measurement for electrical phenomena is none other than the unit of electric current, the ampere, which is very well known among scientists yet little recognized by the general public.

Indeed, while (almost) everyone knows that, between the two terminals of an electrical outlet in their home, there are 110 volts (220 in Europe) and that a cell phone charger charges the battery with a tension of five volts, probably few of us are aware that an ordinary switched-on household appliance consumes a few amperes of electric current or that the circuits of a cell phone carry about one-tenth of an ampere. Yet it is precisely thanks to a measurement of electric current that a differential switch or a circuit breaker protects our household electric systems and sometimes literally saves our lives (hence, in Italian, the name by which a circuit breaker is commonly known, *salvavita*, or life saver). In a not-too-distant future, an electric current of a few million amperes will make a fundamental contribution to the mitigation of global warming.

Scientific Wonders

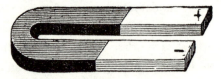

Turning to another subject, I will proceed to explain by what law of nature it comes about that iron can be attracted by that stone which the Greeks call the magnet after the name of its place of origin, the territory of Magnesia. This stone is regarded by people with astonishment; for it often forms a chain of rings suspended from itself.

The passage is taken from book 6 of Titus Lucretius Carus's *De rerum natura* (translated by Martin Ferguson Smith). Already in the first century BCE, Lucretius, a Roman poet and philosopher, recounted how electromagnetic phenomena could put on a spectacle. Similar testimony is also found in the *Naturalis historia* of Pliny the Elder. Confirmation that some manifestations of electricity and magnetism were well known to the ancient Greeks comes from various sources, among them Plato, who wrote in the *Timaeus* (composed around 360 BCE): "And what is more, [this is the explanation for] every kind of water current, even the descent of a thunderbolt as well as that marvelous 'attraction' exercised by amber and by the lodestone" (translated by Donald J. Zeyl). So magnetic stones were known, but it was also known—as we have seen—that a piece of amber rubbed with a wool cloth was able to attract light objects, a phenomenon caused by electrostatic force, which even today is shown to students in physics courses.

It was another few millennia, however, before electromagnetism was understood and its description codified in a coherent theory. In the eighteenth and nineteenth centuries, scientists such as Volta, Ampère, Ørsted, Coulomb, Faraday, Maxwell, and others studied electricity and magnetism, building up to Maxwell's four fundamental equations, which establish the relationships between electric and magnetic fields and their sources, respectively, electric charges and currents.

It is no wonder then that the unit of measurement for electromagnetic quantities arrived late compared to those for length, time, and mass. Scientists started talking about electrical units of measurement in the second half of the nineteenth century, and various proposals were advanced without reaching a general agreement. A major contribution came from the Italian engineer Giovanni Giorgi, born in Lucca in 1871. In 1901, Giorgi presented to the Italian Electro-Technical Association a report entitled *Rational Units of Electromagnetism*, in which he proposed to reform the system by adding to the meter, the second, and the kilogram a fundamental unit of measurement related to electrical phenomena. His proposal was welcomed and marked the start of the difficult task of identifying the unit that was most suitable. Let's not forget that electromagnetism was a young science. While the initial definitions of meter, kilogram, and second were based on thousands of years of experience (recall the measures of length discussed in the chapter on the meter), the choice of the "meters" with which to measure electrical and magnetic phenomena was happening almost in step with their systematic understanding. In 1948, at the Ninth General Conference on Weights and Measures, the unit of measurement for electrical current was chosen as a fundamental unit. This was a crucial step on the way to reaching, in 1960, the definition of the international system.

The definition of the ampere, however, was difficult to put into practice. In substance, its definition was based on the experiments of Ampère, who in turn had been inspired by the Danish physicist Hans Christian Ørsted (1777–1851). However, the story actually began 700 years earlier on the seas of China.

Wires, Compasses, and Currents

Around the second century BCE, when the Great Wall of China was under construction, the Chinese knew that a magnet hanging from a silk thread always pointed in the same direction. The instrument was the precursor of the compass, but at the time, it was used only as a divination tool for predicting the future. It would be a long wait of over a thousand years before it would become a useful aid for orientation and navigation. As Massimo Guarnieri recounts in an article published in *IEEE Industrial Electronics Magazine*, by the early second millennium CE, the compass came to be used first for military purposes on land and later for maritime navigation, which had previously relied only on the stars. The first mention of a compass in Europe dates to 1190 in Alexander Neckam's *De naturis rerum*, and it is still not clear whether it arrived in the old continent from China or was developed independently.

Today, we know that the compass works because the needle made of magnetic material undergoes a force owing to the Earth's magnetic field, which always tends to orient the needle along the north-south axis. Although its effects are very familiar, the origin of the terrestrial magnetic field is not fully known, and much research is being done to explain it. We do know for certain that the terrestrial magnetic field is related to the electric currents flowing in our planet's core of molten metals, but the mechanism with which these currents sustain themselves is still a mystery. Curious readers may want to look on the internet for the images of the computerized simulations by scientists at the University of California, Santa Cruz, which represent the magnetic field inside the Earth as though it were a gigantic bowl of spaghetti!

We are indebted to Hans Christian Ørsted for the experiment that demonstrated that electric currents are at the origin of the magnetic field. Legend has it that, in 1820, as he was conducting a demonstrative lesson on electrical and magnetic phenomena, Ørsted noted with wonder that a compass needle moved when it was placed near a wire carrying electric current. As sometimes happens, the anecdote appears not to be historically accurate. An interesting contribution by Roberto de Andrade Martins to *Nuova Voltiana: Studies on Volta and His Times*, volume 3, recounts that the reality of scientific discovery is sometimes more complex than the way it is simplified in the story. Nevertheless, the fact remains that the observation was surprising, since, up to that moment, electricity (represented here by the wire traversed by the current) and magnetism (the compass needle) were described as two mutually extraneous phenomena. Ørsted, however, demonstrated that it was indeed an electric current that had generated a magnetic field.

News of Ørsted's discovery spread rapidly, and it was Ampère who conducted further and crucial experiments that confirmed and amplified the Dane's discovery and who developed the theory that described them. Ampère then discovered that not only did a magnetic needle undergo a force when it was in proximity to a wire traversed by a current but the same phenomenon also occurred even if the needle was replaced by another wire also traversed by a current. Just so you don't think that all of these developments were just a bunch of academic exercises for specialists in the field, the principle of the force exerted by a magnetic field on an electric current is used in electric motors, for example, in washing machines. If you think about all the science there is behind it, the laundry basket of dirty clothes certainly becomes much more fascinating.

Ampère was able to determine the amount of force between two wires traversed by current as a function of the distance between them. His findings were the basis for the definition of the ampere as a unit of measurement for electric current until 2019. The definition was cumbersome and not very practical: that constant current which, if maintained in two straight parallel conductors of infinite length, of negligible circular cross-section, and placed one meter apart in vacuum, would produce between these conductors a force equal to 2×10^{-7} newton per meter of length. Don't let the definition scare you. There is no need to go into the details. In substance, what this complicated sentence says is the following: take two long wires traversed by the same current, place them at a distance of one meter, and measure the force that attracts them to each other. When this force equals a predefined amount, then those two wires are traversed by one ampere. The predefined amount corresponds precisely to those above-mentioned two-tenths of a millionth of a newton per meter of wire. And there lies the first practical difficulty.

Two-tenths of a millionth of a newton is a very small force. To give you an idea of how small, the weight force of a person who weighs 70 kilograms (154 pounds) (technically, the force of gravity with which he or she is attracted by Earth) is about 700 newtons. Then there is the fact that, according to the definition, the wires must be infinitely long and, above all, that the ampere, despite being an electrical quantity, is defined in mechanical terms, that is, as a force. The unit of measurement of force, the newton, is not fundamental but is derived from the unit of mass in the international system, the kilogram. Finally, we have seen that the value of the prototype kilogram conserved in Sèvres drifted over time and that this value drift limited the accuracy of its derived units. In sum, in both practice and theory, the definition of the ampere in effect until 2019 was not satisfactory. Once again, in order to resolve the problem, we turn to the pillars of nature, or rather to another fundamental constant: the value of the elementary charge (e).

At the beginning of this chapter, we recalled that an atom is made up of protons, electrons, and neutrons. Protons and electrons have the same electric charge, protons with a positive sign and electrons with a negative sign. This electric charge is dubbed elementary. This name derives from the evidence that electric charge is found in nature only in quantities that are exact multiples of the elementary charge, as happens, for example, with eggs. Think of a container full of eggs: a supermarket carton, a wholesale case, a tractor trailer. No matter how many the eggs, they will always be a multiple of the unit. The same is true of the charge. Rub a plastic comb on the sleeve of your wool jacket. Like the ancients' pieces of amber, it will charge and it will be able to attract small pieces of paper. However great the charge deposited on the comb, it will always be an exact multiple of the elementary charge.

The value of the elementary charge is indicated with an e. It is a universal constant, and $e = 1.60217662 \times 10^{-19}$ coulombs. The coulomb is the unit of measurement of the electric charge. In the international system it is a derived unit, hence not fundamental. It owes its name to Charles-Augustin de Coulomb, a French physicist born in 1736, also one of the 72 scientists immortalized on the cornice of the Eiffel Tower. As can be seen from its numerical value, the elementary charge is extremely small compared to a coulomb. To make one coulomb it takes an enormous number of elementary charges, about 6 billion billion of them (or precisely 6.24150907446 × 10^{18}, the exact inverse of 1.60217662 × 10^{-19}). For convenience, let's call it N.

At the beginning of the chapter, we saw that electric current is tied to the movement of electric charges. To be precise, the electric current that traverses a wire is defined as the quantity of electric current—measured in coulombs—that passes through a section of the wire in one second. According to the new definition, approved in 2019, the unit of measurement of an ampere corresponds, therefore, to the passage of N (the enormous number above) elementary charges per second. The ampere, too, therefore, has been liberated from human artifacts (wires, masses, and so on) and has finally been entrusted only to the universal constants of nature, in this case the elementary charge.

Electricity and Sustainable Development

On November 20, 1985, when the two appeared for the closing ceremony, the "personal chemistry was apparent. The easy and relaxed attitude toward each other, the smiles, the sense of purpose, all showed through." The two were Mikhail 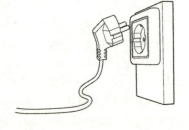 Gorbachev and Ronald Reagan, at the time respectively general secretary of the Soviet Union and president of the United States, who were meeting for the first time at a bilateral summit of the two superpowers. The description was provided by George Shultz, the American secretary of state. The two leaders met at the height of the Cold War to discuss the arms race and especially the possibility of reducing the number of nuclear arms. Held in Geneva, the meeting was the first American-Soviet summit in more than six years, during a period in which the number of nuclear warheads had grown sharply and the strategic relations between the United States and the USSR, as well as the stability of the world, were entrusted to the doctrine of "mutual assured destruction." That doctrine held that if one of the two countries launched a first strike against the other, the second would react and the ensuing nuclear war would destroy them both.

Despite the lack of tangible progress on specific measures regarding nuclear arms, the Geneva summit was a turning point for Soviet-U.S. relations and marked the beginning of the reduction of atomic arsenals, which has continued up to the present (even though the situation is not totally reassuring, since there are still 9,500 nuclear warheads in military stockpiles for potential use).

Beyond the topics related to the arms race, the two heads of state also spoke about the peaceful use of nuclear energy. The official communiqué at the close of the summit noted that the "two leaders emphasized the potential importance of the work aimed at utilizing controlled thermonuclear fusion for peaceful purposes and, in this connection, advocated the widest practicable development of international cooperation in obtaining this source of energy, which is essentially inexhaustible, for the benefit for all mankind." This commitment was soon translated into the start of ITER (in Latin, "the way"), a huge international project for the study of nuclear fusion (the process we discussed at the end of the previous chapter on the kelvin). One year later, an agreement was reached among the European Union, Japan, the Soviet Union, and the United States for the joint design of the program. The People's Republic of China and the Korean Republic signed on to the project in 2003, followed by India in 2005. Despite the good intentions expressed by Gorbachev and Reagan, it took almost twenty years to formalize an executive agreement that allowed the construction of ITER to begin. This achievement speaks volumes about the urgency that the great economic powers have now attributed to the search for carbon-free sources for the production of electrical energy as an alternative to fossil fuels.

Today, the construction of ITER is proceeding apace near Aix-en-Provence, in the south of France, and the first significant results will be seen starting from 2030. The goal of ITER—a reactor that, just to give you an idea of its size, will be as high as a ten-story building—is to demonstrate the scientific and technological feasibility of controlled thermonuclear fusion. ITER will have to produce an amount of thermal power from fusion reactions 10 times as great (500 million watts) as the amount needed to fire up the reactor (50 million watts). ITER will also have the task of laying the groundwork for the next and definitive step: the construction of an experimental reactor, called Demo, able to demonstrate on a large scale the potential production of electricity. If everything goes as planned, Demo will introduce the functionality of fusion in the second half of this century, offering humanity an important tool to fight the environmental crisis.

ITER belongs to the tokamak category, a type of experiment for the study of fusion in a toroidal or doughnut-shaped apparatus, which we discussed in the previous chapter. Its basic working elements are an electric current that flows in plasma—the super-high-temperature ionized gas that is contained in the device and is the fuel for the fusion—and a magnetic field. In essence, in a tokamak reactor, the fusion reactions and their consequent liberation of energy happen in the plasma. It must be heated to a temperature on the order of 150 million kelvin—about 10 times that of the interior of the Sun—and remain confined in a stable and stationary manner inside the reactor without interacting with its metallic walls, something that would drastically decrease its performance. The confinement is obtained by offsetting the expansive force coming from the variation in pressure with an electromagnetic force. The situation is similar to that of an automobile tire. The air inside the tire has a pressure of about two atmospheres, double the ambient pressure on the outside. The confinement of air at high pressure inside the tire is achieved mechanically, that is, by the elastic inner tube. It is the inner tube that exerts a force in opposition to the expansive force originating in the difference between the internal and external pressure of the tire. In the plasma of a tokamak reactor, the situation is similar. The hot plasma in the core of the device is at a higher pressure than the plasma on the outer edges. To contrast the consequent expansive force, a balancing force is needed.

As we have seen in the chapter on the kilogram, Newton's fundamental law ($F = ma$) tells us that if the interaction of a body with its surrounding environment is known, that is, its force F, then we can calculate the acceleration a, which basically means knowing its motion. This equation is also valid in situations of static equilibrium, in which the sum of all the agent forces on the body must be null and, consequently, also its acceleration and its velocity. This is also applied, therefore, to the study of the confinement of plasma, which requires determining a force in opposition to the expansive force of the pressure. That force is obtained by making an electric current flow inside the plasma and, at the same time, applying a magnetic field to the plasma. The mathematics of the solution is relatively simple and is expressed by the elegant equation

$$\nabla p = \vec{J} \times \vec{B}. \tag{1}$$

On the left, the term ∇p indicates the force coming from the pressure of the plasma, which must be balanced by the force originating from the interaction between the electric current that flows in the plasma $\vec{J}$ and the magnetic field $\vec{B}$.

As sometimes happens in science, putting an elegant equation into practice can require a considerable engineering effort, and this is undoubtedly the case with ITER. The current that flows in the plasma, in fact, measures 15 million amperes. By way of comparison, it is over a million times greater than the current that flows in the circuit of an electric oven in an average kitchen. Producing such a current, the necessary magnetic fields, the ultra-high vacuum container that holds the plasma, and a variety of auxiliary components requires cutting-edge technology. The magnetic fields, for example, are produced by magnets that use the principle of superconductivity, which we discussed in the chapter on the kelvin, and building the structure of ITER requires as much steel as it took to build the Eiffel Tower.

A major contribution to the practical realization of fusion will come from this broad international research effort. Important experimental devices are being used to study fusion in the United States, such as the DIII-D tokamak at General Atomics in San Diego, NSTX at the Princeton Plasma Physics Laboratory, HBT at the Columbia University Plasma Laboratory, MST at the University of Wisconsin–Madison, the SPARC device under development thanks to a collaboration between MIT Plasma Science and Fusion Center and the private fusion startup Commonwealth Fusion Systems (CFS), and other devices at additional universities and research centers. In Italy, a new experiment is taking place at the Divertor Tokamak Test (DTT) facility. DTT employs cutting-edge technology conceived in the laboratories of ENEA (Italy's national atomic energy agency) in Frascati and designed by researchers from ENEA, from Italian universities and research centers, and from Eni, Italy's global energy company. With its laboratories in Frascati, Padua, and Milan and many other research centers, Italy is at the forefront of the study of nuclear fusion.

The core of DTT is a steel doughnut about six meters in diameter. Inside its core—caged in by a six-tesla magnetic field, among the highest values ever reached in a large tokamak—the facility will produce a plasma that at its maximum performance will reach a temperature of about 7 million degrees Celsius. DTT's main objective is to be a laboratory of innovation for the study of the intense flows of power released by a fusion reactor. A considerable fraction of the plasma's energy is conveyed, in fact, to a peripheral area of the tokamak known as the divertor. Recent experiments seem to indicate that the power flows that are discharged in the divertor are concentrated on relatively small surfaces with thermal loads per unit of surface area equal to, or even greater than, those on the surface of the Sun. A decidedly "hot" problem for the development of fusion for which DTT will have to find a solution.

That 10 Percent

When asked when fusion energy would be available, Lev Artsimovich, Soviet physicist and fusion pioneer, replied that fusion would be ready when society needed it. Despite its provocative tone, Artsimovich's answer contains a great truth. Starting from the Industrial Revolution, and even more so with the post–World War II economic boom, the rich parts of the world developed and reinforced their own economies using a model based on infinite resources without taking into consideration the consequences for the environment of the increasingly greater amounts of energy derived from fossil fuels. Today, the consequences of those choices are plain for everyone to see, with the dramatic climate crisis and the associated growing awareness that our fossil fuel resources are by no means infinite, as amply demonstrated by all the wars in the Middle East.

The problems demonstrated more violently every day by the climate crisis (and finally echoed by a growing environmental sensibility) have made it clear that a new model of sustainable energy development is urgently needed: a model in which fusion and, in general, research and investment in renewable energy resources and batteries will have a truly important role. Fusion reactions in a future reactor will take place between two hydrogen isotopes, deuterium and tritium. The deuterium contained in a water bottle can generate the same energy as 500 liters of diesel fuel, that is, enough to travel 10,000 kilometers (6,000 miles) in an efficient modern car.

There is, however, another aspect of the energy problem, which we who live in rich countries often tend to overlook: energy poverty.

At the beginning of this chapter, we noted that 90 percent of the world's population has access to electricity. Seven billion people who with a simple gesture—inserting a plug into a wall outlet—can release amperes of current for their electric appliances, their cell phones and car batteries, their heating and air conditioning systems. Not to mention their hospital operating rooms and incubators, the refrigerators that conserve their food, and the pumps that supply their water. A gesture that we take so much for granted that we don't even notice it anymore but that drastically changes the quality of life.

Yet we should not forget about that other 10 percent of the world's population, amounting to about 800 million people, who have no access to electricity, and the additional 2.8 billion human beings who lack access to cooking devices. When we in the developed world want to cook, we simply light our gas or electric ovens. Billions of people, instead, have to use biomass such as wood or dung, and this has serious consequences for their health, since the combustion often takes place in poorly ventilated, closed spaces and is accompanied by particle pollution, which has a dramatic impact on those who spend more of their time in the home, especially women and children. Some 4 million people die each year from air pollution, often caused by these very factors.

The availability of electricity is crucial for our water supply systems. Without electricity, it is not possible to use instruments for the extraction, purification, and distribution of water, which people then have to go get and carry in person. In poor countries, the time needed to fetch water is enormous, and again the burden is borne most often by women and children. According to a study by UNICEF, women in Malawi work for an average of 54 minutes a day to get water, men six. Furthermore, without electricity there is no refrigeration. This means that it is not possible to conserve medicines, vaccines, or food, with tragic consequences. Electricity can make the difference between life and death.

As you are reading this book, masses of desperate people are taking to the seas in rubber rafts or walking for months across hostile and desolate territories, often to be refused entry at border control stations. This should make us think for a moment what it means to have the good fortune to insert a plug into an outlet and how just a few amperes mark, still today, the great divide between the more and less fortunate of our world.

SIX

The Mole

Orange Peels

I was a chemist in a chemical plant, in a chemical laboratory (this too has been narrated), and I stole in order to eat. If you do not begin as a child, learning how to steal is not easy; it had taken me several months before I could repress the moral commandments and acquired the necessary techniques. . . . I stole like [Buck of *The Call of the Wild*] and like the foxes: at every favorable opportunity but with sly cunning and without exposing myself. I stole everything except the bread of my companions.

From the point of view, precisely, of substances that you could steal with profit, that laboratory was virgin territory, waiting to be explored. There was gasoline and alcohol, banal and inconvenient loot: many stole them, at various points in the plant, the offer was high and also the risk, since liquids require receptacles. This is the great problem of packaging, which every experienced chemist knows: and it was well known to God Almighty, who solved it brilliantly, as he is wont to, with cellular membranes, eggshells, the multiple peel of oranges, and our own skins, because after all we too are liquids. Now, at that time, there did not exist polyethylene, which would have suited me perfectly since it is flexible, light, and splendidly impermeable: but it is also a bit too incorruptible, and not by chance God Almighty himself, although he is a master of polymerization, abstained from patenting it: He does not like incorruptible things.

This passage is taken from *The Periodic Table*, by Primo Levi (translated by Raymond Rosenthal), a marvelous book of chemistry and life, considered by the Royal Institution of Great Britain to be the best science book ever written.

In 1937, Levi began his studies in chemistry at the University of Turin. Chemistry is a science that studies matter: how it is made, its structure, its properties, and the transformations of the substances from which it is constituted and how they react. Chemistry is everywhere in our lives and in all of our sensory perceptions: sight, touch, hearing, smell, and taste. Levi—a graduate of a classical high school (*liceo*)—was fascinated by it, and he expressed his fascination in another famous passage from *The Periodic Table*:

> I tried to explain to him some of the ideas that at the time I was confusedly cultivating. That the nobility of Man, acquired in a hundred centuries of trial and error, lay in making himself the conqueror of matter, and that I had enrolled in chemistry because I wanted to remain faithful to this nobility. That conquering matter is to understand it, and understanding matter is necessary to understanding the universe and ourselves: and that therefore Mendeleev's Periodic Table, which just during those weeks we were laboriously learning to unravel, was poetry, loftier and more solemn than all the poetry we had swallowed down in liceo.

Unfortunately, our culture's persistent underestimation of science prevents us from fully appreciating the power of Levi's observation. The periodic table of the elements is, in fact, a sublime construction of human thought, whose foundation dates back to between 1700 and 1800, when chemistry was born in Europe. Despite being the daughter of alchemy, chemistry soon freed itself in that period from the magical aura of its past to become a modern science, thanks to the application of the experimental method, which led to a progressive cataloging and systemizing of the knowledge obtained in the preceding centuries. That early chemistry laid the cornerstone for the development of atomic physics. Chemists discovered new elements, described them, and cataloged them. In the second half of the eighteenth century, Antoine Lavoisier identified 30, most of which are also elements according to the current definition, but already by the end of the nineteenth century, the number of known elements had risen to 70. Today, we know 118: 92 of them natural and the rest obtained artificially.

The turning point came about thanks to the Russian Dmitri Mendeleev (1834–1907), who arranged the elements in relation to each other in a simple and schematic way, the periodic table, which he proposed in 1869. His system was as simple as it was ingenious: he arranged the elements in rows and columns based on their atomic weight. This classification was later improved by using the atomic number, or rather the number of protons that are part of the specific element. As in the assembly of a jigsaw puzzle when the initial confusion of a pile of pieces morphs into an emerging image, the arrangement of the elements in that way unveiled unexpected relationships among themselves and, above all, prodded chemists to keep on searching. Mendeleev, in fact, was not afraid of the empty spaces in his table—just the opposite. Like the great scientists, he considered doubt and ignorance not as shameful but as a resource. He hypothesized that where there was a missing element, with its corresponding atomic weight, a space should be left empty. The empty space suggested that the element that belonged there hadn't yet been discovered. His classification was a fundamental contribution to chemistry, illustrating new properties of the elements and demonstrating that they present themselves in different forms and combinations, but always in a finite number. Today, the elements of the periodic table are arranged according to their atomic number, thanks to the work of Antonius van den Broek from 1909 to 1913.

Let's get back to Primo Levi. Despite the Italian racial laws that instituted compulsory discrimination against Jews, he took his degree in 1941 and found work. In 1942, he joined the clandestine Action Party. After September 8, 1943, when Italy signed the armistice with the Allies, he joined a group of partisans operating in Valle d'Aosta, in Italy's northwest. A few months later, on December 13, in Brusson, the fascists, who did not know about his activity in the Resistance but who identified him as a Jew, captured him. Taken to the concentration camp in Fossoli, he was then deported to the Auschwitz-Birkenau extermination camp.

Levi's chemistry degree, along with the little German he had studied to read some of his university textbooks, made him a useful prisoner and probably saved his life. "*With these empty faces* of ours, these shaved skulls, these shameful clothes," he recounts in *If This Is a Man* (translated by Stuart Woolf), he was given a "chemistry exam" before a Nazi officer. He was recruited into the Kommando 98, a unit of specialists also called the Chemical Kommando, and then put to work in a chemical factory not far from the camp. It was a very lucky assignment, not least because there he was able to find a lot of things to steal and exchange for food rations on the black market in the concentration camp, as he recounts in the passage with which we opened this chapter. Levi managed to survive until the liberation of the camp in January 1945.

Moplen!

For Primo Levi, one bottle would have been enough. In 2021, according to the authoritative website Statista, we produced worldwide

583 billion plastic bottles. This number trans-
lates into 49 billion bottles a month, 1.6 billion per day, 67 million per hour, and about
1 million per minute. If we were to pile
them up, one on top of another, at the same
rate of production, it would take one minute
to make a column that reached . . . the International Space Station.

All these bottles amount to an enormous quantity of plastic, adding to that which is manufactured for other purposes. The cumula-
tive production from the 1950s to today amounts to 8.5 billion tons
of plastic, most of which is still with us. With plastic, for the first
time in human history, we have produced and consumed on a global
scale a material that lasts much longer than we do because it takes
such a long time to biodegrade. The only hope is to recycle it, something we started to do only recently and still in minimal amounts.
As a result, since we started to mass-produce it after World War II,
plastic has accumulated: more than 6 billion of those 8.5 billion
tons are scattered over land and sea, polluting our planet. As reported by *National Geographic Magazine*, the oceans contain an estimated 5.250 trillion pieces of plastic detritus, most of which do not
float but sink down to the depths, with devastating consequences
for the environment. A dramatic emergency, which we are only now
starting to recognize but that Primo Levi prophetically perceived
as early as 1975, when he wrote *The Periodic Table*.

Those were the years when the world was falling in love with
plastic. Light, resistant, colorful, durable—even too much so, in
hindsight—plastic became one of the symbols of modernity and the
economic boom. The Italian chemical industry played a starring
role in the development of this new material. Giulio Natta invented
isotactic polypropylene and won the Nobel Prize in 1953. Isotactic
polypropylene, a scary name that conjures images of mad, white-
coated laboratory scientists. A world away from the impression
created when it was called by its commercial name—Moplen—

pronounced with a toothy grin in Italian television commercials by the great comic actor Gino Bramieri (1928–1996). Throughout the 1960s, ads for plastic were a regular feature of *Carosello*, the ten-minute advertising slot screened every evening at around 8:45 p.m. The smiling Bramieri, surrounded by plastic buckets, colanders, coffee cups, and toy cars, crooned, "Mo-mo-moplen!" That's right, Moplen was the cheery, comforting nickname invented by Italy's advertising "mad men" to sell the isotactic polypropylene that revolutionized our homes in the sixties' economic boom.

In the concentration camps, lightweight and splendidly impermeable polyethylene, the most common among the plastics, would have been very useful for Primo Levi. Especially polyethylene terephthalate, or PET, a thermoplastic widely used in food packaging, bottles included. Which, however, with characteristic subtle irony, Levi describes as "a bit too incorruptible." In nature, it takes hundreds of years for it to biodegrade.

A Bad and Undeserved Reputation

Like the adjective *nuclear*, the noun *plastic* now has a terrible reputation. Demonizing nuclei and plastics makes us feel better, but it is unlikely to lead to a solution for complex problems. Nuclear medicine, for example, is an indispensable achievement of modern medicine, and nuclear energy will perforce be a  necessary component of a basket of sustainable, carbon-free energy resources. Similarly, a multitude of plastic items have changed the quality of our lives for the better. Think of how plastic is used in hospitals: syringes, intravenous bags, catheters, scalpels, just to give a few examples—and how it has improved the hygiene and efficacy of medical treatment. Think of its use in the sterile conservation of medicines and foods; in the production of helmets for bikers and cyclists, infant seats, airbags; in reducing the weight of cars and other means of transport, thus reducing their fuel consumption and consequent emissions of CO_2. The problem is not plastic but our "use once and throw away" culture, a clear contradiction in terms compared to the eternal life of the material. The fault lies not with plastic but with, for example, the practice of packaging prepeeled oranges to be sold in supermarkets, or producing untold quantities of water bottles that are used and thrown away in a few minutes' time, or using redundant packaging, or not reusing receptacles and containers.

The responsibility is ours, especially those of us who live in the rich part of the world. Every day, our actions as human beings have an impact on the Earth and our environment. Some of those actions are useful. Others are not. We can save ourselves only with responsible individual and collective behavior, and with science. Science, and especially chemistry, can make an enormous contribution to a cleaner, more sustainable world. First, through science education and popularization. Responsible behavior and sustainable public policies and practices prevail where there is more and reliable information—as opposed to hearsay and urban legends—and thus more awareness of problems and their possible solutions. Then, naturally, through research.

Knowing is important. Consider, for example, the chemical formula that describes the combustion of methane:

$$CH_4 + 2O_2 \rightarrow CO_2 + 2H_2O. \tag{1}$$

It looks complicated, but if we read it carefully, it reveals a lot of information. Combustion is a rapid chemical process of oxidation, during which the fuel reacts with an oxidizing agent, or oxidant, called a comburent. Through oxidation, the oxidized substance loses electrons to the oxidizing substance. In the case of combustion, the oxidant is typically oxygen, while the fuel may be gas, liquid, or solid, either natural or artificial. Chemical energy stored in the fuel is transformed into thermal energy (the heat associated with the flame) and, often, also into electromagnetic radiation (light).

Getting into detail, equation (1) tells us that a molecule of methane (CH_4) is composed of one atom of carbon (C) and four atoms of hydrogen (H). It also confirms that the combustion of methane requires oxygen (O_2). Who doesn't remember the experiment we did as kids, with the candle that goes out when you put a glass over the flame and all the oxygen is consumed? When combustion takes place with insufficient oxygen, it also produces the dangerous and poisonous carbon monoxide (CO).

Finally, the formula makes it unequivocally clear (it's a formula, not a rumor) that when methane is burned, it produces carbon dioxide (CO_2) and water (H_2O). There lies the problem. Methane is a fossil fuel, and like all fossil fuels—coal, gas, petroleum—it contains carbon. The adjective *fossil*, in fact, derives from the Latin *fossilis*, that which is obtained by digging, which, in turn, derives from the verb *fodere* (dig). The term *fossil* refers generically to any residue of a vegetable or animal organism that lived in the past and whose remains have been enveloped in the Earth's crust. The current fuels in our world are, in fact, prehistoric plants and animals that lived hundreds of millions of years ago. When they died, they were buried under layers of rock, mud, and sand, over which there sometimes formed bodies of water. Over millions of years, they decomposed and formed fossil fuels. Oil and gas, for example, originate from organisms that lived in water as algae and protoplankton.

When any fossil or biological-origin fuel burns, it produces CO_2 and thus contributes to the greenhouse effect. A simple formula, made up of a few symbols, explains elegantly and inexorably what awaits us if we fail to change our ways very quickly. Note well: what awaits us human beings. Because the planet will survive our misdeeds, and the luxuriant forests around Chernobyl are there to remind us of that. We are the ones who risk disappearing.

As we have seen at the beginning of this chapter, chemistry is part of the world that surrounds us, and it helps us to utilize the planet's resources for our well-being. Chemists design many of the materials that compose the objects that we use each day, from electronics to medicine, and they help humans to use the resources that the planet places at our disposal. Yet chemistry also has an important role to play in improving the overall sustainability of our world and in satisfying the needs of its inhabitants, especially of those who inhabit its poorer regions, as well as future generations.

Green chemistry, as it is called, or the science of sustainability, can help not only to clean up the planet but also to avoid pollution. It can help us understand, monitor, protect, and improve the environment that surrounds us. For example, by developing instruments and techniques for observing, measuring, and reducing air and water pollution. The detailed understanding of pollutants is crucial for understanding their effects on health—for example, the correlations between health problems and atmospheric pollution—and for developing technologies for their reduction. Accurate measurement makes it possible to verify the observance of policy measures for the improvement of air quality. Science also makes it possible to reduce exhaust emissions by developing cleaner fuels, increasing engine efficiency, inventing new technologies for automobiles—such as batteries and fuel cells for hydrogen vehicles—and improving devices for controlling vehicle exhaust pollution such as absorbers, antiparticulate filters, and three-way catalytic converters that reduce carbon dioxide, unburned hydrocarbons, and nitrogen oxides in gasoline engines. Perhaps soon, even clothes and buildings will be able to purify the air by way of photocatalytic processes using only oxygen and light.

Posthumous Fame

Among the mythical 72 inscribed on the Eiffel Tower is one who almost didn't make it: Gustave de Coriolis, who gave his name to an

important physical phenomenon known as the Coriolis
effect. The Coriolis force is
observable in a body moving
in a rotating frame of reference like the Earth. It is responsible, for example, for the
formation of cyclonic and an-
ticyclonic weather systems in
the atmosphere, and it is im-

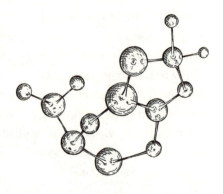

portant in ballistics. A force that would be lacking if the Earth did
not rotate. Giovanni Battista Riccioli (1598–1671), Jesuit and Ptole-
maic astronomer, intuited its existence, but because he still did not
have the instruments to measure it and, above all, because he was in
some way a victim of his own geocentric prejudice, he concluded
that since (in his view) the Earth did not move, his theory must
have been mistaken. In so doing, he left the honor of the discovery
and the fame to Coriolis 150 years later, contenting himself with
a lesser place in the history of science, thanks to his drawing one
of the first lunar maps and to having an asteroid and a previously
unknown Moon crater named after him.

Someone who got it right but was misunderstood by his contem-
poraries was the Italian scientist Lorenzo Romano Amedeo Carlo
Avogadro, Count of Quaregna and Cerreto. Amedeo to his friends,
fellow scientists, and students. Born in Turin in 1776, he studied
law, specializing in canon law. It turned out that codes and pandects
were not his cup of tea, so he shifted his interest to science, rapidly
attaining brilliant results that were to become pillars of modern
chemistry. In particular, he outlined the law that today bears his
name. It states that equal volumes of different gases, at the same
temperature and pressure, contain the same number of molecules.
The same finding was made some years later by André-Marie Ampère, whom we discussed extensively in the preceding chapter.

In his writings, Avogadro introduced the distinction between "elementary molecules" and "compound molecules," and he hypothesized the possibility of the division of compound molecules. As Marco Ciardi notes in his book *Il segreto degli elementi* (The Secrets of the Elements), Avogadro's elementary molecules were, at the time, thought of not as real physical entities but rather as abstract entities of a mathematical nature. These were extremely innovative concepts for the time, which were ignored by the scientific community, despite providing one coherent explanation of otherwise incomprehensible experimental observations. Only in 1860, four years after Avogadro's death, and thanks to the contribution of another Italian chemist, Stanislao Cannizzaro, was his theory's fundamental value finally recognized, leading, over the span of a few decades, to the identification of "elementary molecules" with atoms and "compound molecules" with molecules in today's parlance.

The scant recognition that Avogadro enjoyed in life has been compensated by giving his name to what has become the most important universal constant in chemistry. Avogadro's number has today made possible the redefinition of one of the seven fundamental units of the international system, the mole.

You Want to Be a Millionaire?

Understanding large numbers is not generally an easy task for the human mind, nor is it very practical. While we can manage to visualize a dozen people, a couple of dozen sheep, a hundred and some books, when the numbers start getting bigger, we all start to lose our bearings. Think, for example, of the definition of millionaires and billionaires. Sure, all of us would rather be a billionaire than a millionaire, because billions are more than millions. But how much more? Is there really all that much difference?

Intuitively, it's hard to say, not least because most of us have no experience of what it means to be a millionaire or a billionaire. Indeed, according to an analysis of the Boston Consulting Group, there are about 400,000 people in Italy with an investible capital greater than $1 million, about 1 percent of the adult population. Not very many. If, on the other hand, we transform the million and the billion into something with which we have a little more direct experience, like time, then things change. Let's take a million seconds, which add up to about 11.5 days. A billion seconds, on the other hand, correspond to 32 years. A Christmas vacation versus a significant portion of an entire life. Now that's a difference we can all understand.

Big numbers are often also impractical. Let's suppose that a printing shop has to print 10,000 copies of an 8.5 × 11-inch flyer. At the warehouse, the printers check on the paper supply. If they were to count the paper one sheet at a time, it would be an inefficient task. Much easier to count the reams, knowing that each ream contains 500 sheets. If they count at least 20 reams on the shelves, the operation is done in the blink of an eye. They're set to go. If there are fewer, the printers have to make an order. A similar logic applies to fruit. We buy watermelons by the unit, but we certainly don't do that with cherries, which instead we buy and sell by weight. Nobody would ever think of buying 100 cherries, and we'd have no idea if they were too many or too few for our pantry. But if we think of 800 grams (1.75 pounds) of cherries, the approximate weight of 100 cherries, we realize immediately that they are just right if we serve them as part of a dinner for six guests, though they are decidedly too many if we want to eat them as a snack just for ourselves.

A similar thing happens when you study the microscopic world or when you penetrate down into the meanderings of matter, as physicists and chemists do. An example? Water. In chemical notation, its molecule is written H_2O. It is composed, that is, of two atoms of hydrogen and one atom of oxygen. Producing it involves making hydrogen and oxygen react in the right proportions, in accordance with a chemical formula that chemists write as

$$2H_2 + O_2 \rightarrow 2H_2O. \tag{2}$$

Translated from the language of chemistry into English, the formula reads like this: two molecules of hydrogen combine with one molecule of oxygen to create two molecules of water. These proportions need to be noted: the molecule of oxygen has two atoms, while in a molecule of water one oxygen atom is enough. Now let's say that we want to produce a certain amount of water, starting from oxygen and hydrogen. If we want to make homemade tagliatelle noodles, all we need to do is use one egg for every 100 grams of flour. It's relatively easy to count the eggs and weigh the flour. But with atoms and molecules, what do you do? They are microscopic objects—an atom of hydrogen is smaller than a billionth of a meter—practically invisible. They certainly can't be counted by way of direct observation. We must try to convert the measure into something more manageable. This is where the mole comes into play; in fact, it's a sort of ream.

To measure the number of microscopic elementary units (atoms or molecules), you use a physical quantity called amount of substance. The amount of substance, or of matter, is one of the seven fundamental physical quantities of the international system. It is a fundamental quantity for chemistry and measures how many atoms or molecules are present in a given amount of matter, for example, how many molecules are present in a liter of water. Its unit of measurement is the mole, and here is where Avogadro comes in. Just as it was decided that a ream of paper contains exactly 500 sheets, it has been established that a mole contains exactly a number of elementary entities of a given substance equal to . . . are you ready for a big number? A really big number? Try 602,214,076,000,000,000,000,000. A huge and hugely important number for science; a universal constant named especially after Avogadro in recognition of his essential contribution to the development of modern chemistry. Avogadro's number emerged from heated scientific debate and was calculated with increasing precision until, with the revolution of the international system in 2019, it was adopted as a fundamental constant for the new definition of the mole, which had previously been determined in a much more cumbersome way and was dependent on the kilogram.

A mole of substance is something much more manageable and, in terms of weight, macroscopic. A mole of molecular oxygen corresponds to 16 grams of substance. A mole of molecular hydrogen corresponds to two grams. A mole of water corresponds to 18 grams. These numbers are much more reasonable compared to the mass of a single molecule of hydrogen, which, expressed in kilograms, is a number so small that it has 26 zeroes following the decimal point. The mole, therefore, connects the invisible microscopic world to the observable macroscopic world and, like the ream for printers, makes the work of chemists much more practicable. Thanks to the mole, it is easy to make a connection between the amount of a substance and its mass in kilograms or grams. Avogadro's number is a conversion factor, a sort of bridge, that transfers the "frightening" numbers of atoms and molecules into the more calming numbers of amount of substance.

Those Who Do Not Look the Other Way

It is November 9, 1943, and the University of Padua is inaugurating its 722nd academic year. Italy is going through one of the darkest periods of its recent history. Just a couple of months earlier, on September 8, was the announcement of the armistice with the Allies, followed by Mussolini's liberation by the Germans from his imprisonment in Abruzzo and the proclamation of the fascist Italian Social Republic. The recently nominated rector of the university is Concetto Marchesi, a renowned Latinist but also a member of the Italian Communist Party and a proud antifascist. Marchesi had taken office in early September, appointed by Mussolini's replacement, Prime Minister Pietro Badoglio, to take the place of his predecessor, chosen by the fascist regime. His academic intent was crystal clear, and he synthesized it in an interview on September 10 with the newspaper *Il Messaggero*, which ends like this:

> [This] new life must start pulsating right away in Italian universities. . . . My purpose is to promote forthwith the free intellectual training ground of university studies . . . where it is possible to discuss and experience what freedom is, what are the economic and political doctrines that are to be welcomed or rejected, what, finally, are the supreme interests of the Nation, of the people, of working people. This must be the new air that penetrates Italian universities right now, this the new breath that must be granted right away to the young people of our universities.

Soon, however, the situation worsens and Northern Italy is crushed by the reborn fascism of the Republic of Salò, as Mussolini's regime was popularly known, from the village of Salò on Lake Garda, where some of the Social Republic ministries were moved. Marchesi presents his resignation as rector, which is not accepted by the Italian Social Republic. His academic reputation was such that, in a unique twist of fate, one of the greatest universities of Northern Italy found itself with a communist rector. The November 9 inauguration, therefore, becomes an unmistakable political act against the regime. As the newspaper accounts tell it, a minority of students wearing the uniform of the fascist republic storm the overflowing auditorium, attempting to force the inauguration to profess fidelity to Mussolini and to the Italian Social Republic. Marchesi himself gets involved in expelling them from the ceremony and gives a speech that will go down in history.

"There is something new or unusual," Marchesi begins, "like a great sorrow and a great hope, which brings us together to listen, not so much to the fleeting words of a man, but to the centuries-old voice of this glorious university, which today calls the roll of its teachers and disciples; and the teachers and the disciples here present answer for the distant, for the missing, for the fallen. So, in this small circle, among us, today, a rite is performed that makes our suffering sacred and our hope secure." The rector then goes on, "The university is certainly the highest intellectual training ground for young people: where, slowly or impetuously, the problems come to the fore, where minds are more intent on knowing or recognizing those that will, perhaps, remain the fundamental truths of individual existence. Moreover, we teachers have the duty to reveal ourselves wholly, without closure or reticence, to these young people, who ask us not only what are the aims and procedures of the particular sciences, but what stirs in this yet vast and infinite and mysterious journey of human history."

Marchesi concludes his speech with a heartfelt appeal (for the full text, see the Universitá di Padova website, https://800anniunipd.it /en/storia/il-discorso-di-concetto-marchesi-dinaugurazione-del -722-anno-accademico/):

> Gentlemen, in these hours of anguish, amidst the ruins of an implacable war, the academic year of our University reopens. That none of us, dear young people, should lack the spirit of salvation. When it is with us, everything will rise again, that which was wrongly destroyed, and everything will be accomplished, that which was rightly hoped for. Young people, trust in Italy. Trust in its good fortune, as long as it is sustained by your discipline and your courage: trust in the Italy that must continue to live for the joy and the honor of the world, in the Italy that cannot fall into servitude without plunging into darkness the civilization of peoples. On this day 9 November 1943 in the name of this Italy of workers, artists, and scientists, I declare open the 722nd year of the University of Padua.

That impassioned speech for freedom and against the servitude of fascism that would have plunged civilization into darkness marked a turning point. A few weeks afterward, Marchesi had to leave his home for the shelter of friends in Padua and then in Milan. From there he had to flee to Switzerland, where he remained until the end of the conflict, keeping in close contact with the Resistance, one of whose leaders in the Veneto region was Egidio Meneghetti, professor of medicine and pharmacology and vice-rector in Padua under Marchesi until 1943. An eminent scientist, he made fundamental contributions to medicine. Starting in 1943, he participated actively in the Resistance. Arrested and tortured in 1945, Meneghetti was transferred to the concentration camp in Bolzano pending deportation to the extermination camps, from which he was saved only because the intense Allied bombing in the last months of the war disrupted the railroad lines from Northern Italy to Germany.

A little over a century earlier, Amadeo Avogadro held the chair in sublime physics at the University of Turin, dedicated to the mathematical principles of the sciences. (If today this definition of physics may prompt a smile, let's not forget its origins: *sublime* comes from the Latin *sublimis*—with the variant *sublimus*—a compound of *sub* [under] and *limen* [lintel or threshold], "reaching to the upper limit," meaning lofty or distinguished.) In 1820–21, Avogadro was close to the leaders of the revolutionary uprisings that shook Europe and had repercussions in the Piedmontese academic world and its student movement. For this reason, King Charles Felix suppressed several university professorships in 1822, including the chair held by Avogadro. Indeed, the university was "happy to allow this interesting scientist to take a rest period from the heavy burden of his teaching duties, so that he is able to better attend to his research."

Avogadro did not turn the other way, much like Marchesi, Meneghetti, Fritz Strassmann, Silvio Trentin, and many others in the times of Nazi-Fascism—the list could, unfortunately, be very long indeed—in countries where even today freedom of thought and academic freedom are in danger. These were humanists, scientists, and physicians who made the critical thinking inherent in study the measure of their being in society.

The motto of the University of Padua, which in 2022 celebrated its eight hundredth anniversary, is "Universa Universis Patavina Libertas" (Paduan Freedom Is Universal for Everyone). A universal aspiration that universities and places dedicated to the pursuit of knowledge and education will always be beacons of liberty, welcome, and tolerance.

SEVEN

The Candela

Portrait of the Ninth Day

 The portrait of Ferdinando II de' Medici, painted in 1626 by Justus Sustermans and conserved in the Palatine Gallery of the Pitti Palace in Florence, is the least "Botticellian" face in the collections of the Uffizi Galleries. That was the definition given by the Uffizi in a post published on its Facebook page on February 9, 2021, and it is immediately understandable as soon as you see the painting. Sustermans, a famous portraitist, court painter, and author of two celebrated portraits of Galileo (also conserved in Florence), fixed on canvas the then 16-year-old Ferdinando II, ill with smallpox, on the ninth day of infection. The face of the young nobleman is covered with cutaneous pustules, the classic symptoms of the disease. Pustules would also form on the sick person's torso, accompanied by a

high fever, and they would invade the oropharynx, precluding the patient from eating, often with fatal outcomes. Those lucky enough to survive were plagued with deep facial scars for the rest of their lives.

That portrait may have influenced Leopold II, grand duke of Tuscany, in his choice to be "pro-vax" before the term was coined. In 1786, the grand duke entrusted first himself and then his children to the Dutch scientist Jan Ingenhousz as test subjects for the pioneering technique of smallpox immunization, so-called variolation, of which Ingenhousz was an expert practitioner. The technique consisted of making a superficial incision on the patient's skin with a needle dipped in the pus of a smallpox pustule taken from a sick person. Ingenhousz, born in 1730, had tried it with success in England on hundreds of people. Thanks to his fame, he was called by Empress Maria Theresa of Austria, who asked for inoculations for herself and her family. Smallpox was a rather terrible disease, believed to have killed some 60 million people in Europe. In his *Letters on the English*, a series of essays published in 1733 and inspired by his sojourn in England, Voltaire wrote that the incidence of smallpox infection approached 60 percent of the population, with a mortality rate as high as 20 percent. He also expressed his wish that the practice of inoculation would take hold in France as it had in England.

A few decades later, Edward Jenner developed the first proper vaccine, also against smallpox, which was also the first highly effective vaccine developed worldwide. Jenner noticed that farmers who came into contact with cowpox pus were often immune to the virus that affected humans, or at least their symptoms were much more benign. Cowpox produced pustules similar to those caused by smallpox. So Jenner decided to inoculate, not using the human virus, as had been done up to that time with variolation (obviously with a certain risk), but rather using the virus that affected cows, *Vaccae* in Latin, from which comes the name of the technique, vaccination.

Thanks to the vaccine, smallpox, a disease whose first clinical traces—present in Egyptian mummies—date back to 3,000 years ago, has been completely eradicated.

Today, some 200 years after Voltaire's *Letters on the English*, there are still some who do not believe in vaccines. Learning from history is not for everyone.

Sugars and Oxygen

The history of science has not been kind to Jan Ingenhousz. He is not known by many, although he did acquire a bit of notoriety thanks to the "doodle" (the modified version of its logo) that Google dedicated to him on December 8, 2017, the 286th anniversary of his birthday. Yet Ingenhousz certainly deserves an important place in our collective memory. In 1779, he made a fundamental contribution to our understanding of photosynthesis, the process by which plants convert sunlight into chemical energy.

A few years earlier, Joseph Priestley conducted experiments that demonstrated how a plant was able to regenerate the oxygen consumed by a candle that burned out under a closed bell jar. Ingenhousz deserves credit for having understood the decisive role of light in association with plants. He noted that leaves produce oxygen in sunlight and carbon dioxide in darkness. He published these findings in 1779, thus exercising a decisive influence on further research into vegetable life in the centuries that followed.

Today, we know that plants, algae, and Cyanobacteria use sunlight, water, and carbon dioxide to create oxygen and store energy in the form of sugar molecules. Photosynthesis is a crucial process for life on Earth because it enables the collection and transformation of an enormous amount of solar energy. Most living beings rely on photosynthesis to produce the complex organic molecules they rely on as an energy source. The sugars produced during photosynthesis are the base for more complex molecules obtained from the photosynthetic cell, such as glucose. Consider that, on average, photosynthesis on Earth uses about 30 trillion watts, about five or six times the power demand created by all human activities.

Beyond the transformation of energy, another effect of photosynthesis, also fundamental for life, is the release of oxygen into the terrestrial atmosphere. Most photosynthetic organisms generate oxygen as a by-product of this process, and the advent of photosynthesis changed life on Earth forever. Photosynthetic organisms also remove great quantities of carbon dioxide from the atmosphere and use the carbon atoms to construct organic molecules.

Blue Water, Clear Water

One of the astounding coincidences that made life possible on Earth is the incredible interaction between sunlight and water. Two totally different and physically independent entities whose properties, however, are strongly interconnected.

The Sun, our private star, is a natural nuclear reactor inside of which, thanks to fusion, enormous quantities of nuclear energy are converted into other forms of energy, some of which, such as electromagnetic radiation, make it to Earth. The Sun fuses 600 million tons of hydrogen per second and irradiates onto the most external parts of the Earth's atmosphere something like 1,360 watts per square meter, a number known as the solar constant. A very large quantity, if you consider that a square meter is more or less the surface area of a kitchen table. If we could capture and reuse all the solar energy that arrives on a square meter on the edge of the atmosphere in just one hour, we could power a refrigerator for an entire day.

Considering the spherical form of the Earth, and that at any given time only a part of its surface is exposed to sunlight, on average the Earth receives about 340 watts per square meter. All of this energy enters into a delicate environment whose major factor is water. Water vapor is, in fact, one of the primary actors in the greenhouse effect, the natural process—today, dramatically influenced by humanity—that ensures that the Earth is not a frozen sphere wandering around the universe but rather the cradle of innumerable forms of life.

The average temperature on the surface of our planet is 14°C (57.2°F). This temperature is maintained by solar radiation, a third of which is reflected by the Earth while two-thirds is absorbed. The absorbed portion is in turn reemitted, still in the form of electromagnetic radiation, but with a different aspect. The electromagnetic energy radiated by the Earth has different frequencies from those of the energy coming originally from the Sun, and it is mainly in the infrared range. Not a trivial difference, because the infrared range is absorbed by the atmosphere and radiated back to Earth again. This process amplifies solar warming: it is the greenhouse effect. Today, this effect has a negative reputation, but in reality, in its natural form it is indispensable for life on Earth. Without the greenhouse effect, the average temperature of the planet would plummet from the current 14°C to –18°C (–0.4°F).

The creators of the greenhouse effect are gases that make up only a small part of the atmosphere. The gases comprising most of the atmosphere, such as nitrogen and oxygen, which respectively account for 78 and 21 percent of the atmosphere, produce a negligible greenhouse effect. Instead, the main actor is water, or rather water vapor, which makes up about 1 percent of the atmosphere. Equally important, though present in even lower concentrations, are carbon dioxide and methane. The former, also known as CO_2, has an atmospheric concentration of about 400 parts per million. Methane's concentration is even lower. Despite its low concentration, CO_2 plays a crucial role as a regulator. Its variation, in fact, modifies the atmospheric temperature, which, in turn, changes the water vapor content, which has a much more dramatic impact on the greenhouse effect. It is, as you can see, a delicate balance, where small variations cause large effects. That is why the continuous increase in the atmospheric concentration of CO_2, caused by humanity, is so dangerous for our planet.

The interaction between sunlight and water also takes place in the oceans. Of all the electromagnetic radiation emitted by the Sun, only certain wavelengths are perceptible to the human eye. Normally, sunlight looks yellowish white, but actually it is composed of many colors "mixed" together. This is the so-called visible spectrum, characterized by wavelengths measuring between 400 and 700 billionths of a meter. Each wavelength interval corresponds more or less to one color (technically this statement is not all that precise, given that the solar spectrum is continuous and so precise demarcations between colors cannot be made). The single colors are visible, for example, in the rainbow. A basic, and at the same time extraordinary, fact for life on Earth is that water has a preference for visible light.

In general, water is an excellent absorber of electromagnetic radiation. There are multiple practical examples of this phenomenon. One that nearly all of us observe in our homes is the process of cooking in a microwave oven, which happens precisely because the water molecules contained in the food absorb that specific wavelength in the microwave spectrum of electromagnetic radiation. Other, more technological, examples are the difficult radio communications between submarines or the spent fuel pools where exhausted nuclear fuel rods from fission power stations are stored, since water is also an excellent absorber of high-energy radiation.

Yet this absorbent behavior has a noteworthy exception: visible light. In the narrow wavelength interval between 400 and 700 billionths of a meter—a trifle with respect to the entire spectrum, which goes from millionths of billionths of a meter up to tens of meters—water transmits light. It is transparent. That interval is exactly the interval of visible light. It is the interval in which solar radiation has its peak, to which the eyes of humans and animals are sensitive; the interval that is absorbed by plants and algae in photosynthesis. In the ultraviolet, just above the visible, the rate of absorption by water increases steeply, reinforcing our protection against the ultraviolet rays of the Sun.

The transparency of water is a key factor in marine ecology. Thanks to light, marine animals can see their prey. The Sun is also a fundamental energy source for all biological phenomena, and the penetration of sunlight gives rise to photosynthesis, which produces nutrients for marine fauna and, as a result, for the planet. We have to thank the oceans for every breath we take, since it is estimated that they produce—by way of the photosynthesis carried out by algae and phytoplankton—at least half of the oxygen in the atmosphere. This process has been going on in the oceans since long before it was carried out by land-based plants, whose oldest fossils date to about 470 million years ago. Fossils of Cyanobacteria and algae, by contrast, date back as far as 3.5 billion years ago.

The spectrum of water absorption and the peak of solar radiation are obviously physically independent phenomena, but it is a great good fortune for life on Earth that they coincide.

To the Measure of Man

Expressions describing important transitions such as the emergence of something new ("coming to light") or the end of our lives ("closing our eyes forever") indicate how important light and vision are for human life. Vision may be the most powerful of the human senses, and light is an omnipresent symbol in the great religions, as well as one of humanity's ancient metaphors for discernment, wisdom, and truth. In the Bible, for example, right after creating Heaven and Earth, God creates light (Gen. 1:3), even before the stars. No wonder then that the unit of measurement for light, or more precisely for luminous intensity, is the unit of the international system most tied to human beings, and even today, in our hypertechnological world, it bears the name of one of the oldest instruments of illumination, the candela (Latin for "candle").

The candela is the seventh and last base unit of the international system. It measures luminous intensity, which, if we want to use a technical term, is the power emitted by a punctiform luminous source per unit of solid angle in a given direction. This is a rather esoteric definition, but we need not go into detail. The important thing is that the candela is the base unit for photometry, the science of the measurement of the light perceived by the human visual system. It does not represent the whole amount of light emitted by an object; that is described by the radiant flux, and it has a unit of its own, the lumen, known commonly as the measure that appears on light-bulb packaging. This means that a light bulb's performance is not measured in candelas but in lumens, which indicate the total amount of light emitted in all directions. The lumen responds to a practical need because it tells us how well the bulb will illuminate the entire environment in which it is located.

The candela, instead, is a measurement of the luminosity of a light source when it is looked at directly. To describe mathematically the idea of the light that radiates in three-dimensional space, we use the solid angle, which is essentially the three-dimensional projection of a planar angle. Take a pie and divide it into six slices. Each slice defines a 60° angle with its vertex at the center of the pie. Now consider a ball. A solid angle is a cone-shaped section of a sphere with its vertex at the center of the ball, and it is measured in steradians. So, while the lumen measures all the emitted light, the candela measures only the light an observer looking at the light bulb can see. Moreover, the candela is not used for any kind of electromagnetic radiation, such as, for example, X-rays, microwaves, radio waves . . . , which are measured in watts. The candela, therefore, is a rather peculiar and very anthropocentric unit. It measures the light to which our eyes are sensitive—visible light—coming directly from a source that we can see, and specifically the part of it that ends up in our eyes.

For millennia, the flame was our only artificial light source, and it still was in 1875, when 17 nations signed the Metre Convention in Paris. The development of electric light bulbs had only just begun, and in 1878, Joseph Wilson Swan patented the incandescent light bulb with a carbon filament. Contemporaneously, in the United States, Thomas Edison was working away, and three years later, in 1881, the Savoy Theatre in London became the first public building to use incandescent light bulbs. It would take another decade, however, before electric light became widespread.

The watershed year was 1948. Up to then, various standards were used to measure luminous intensity. In general, they were based on the luminosity of a real candle flame as a reference, which had a well-defined composition and form, or on the luminosity of an incandescent filament, it, too, with well-defined properties. Inevitably, the various units were not uniform, and as also happened for the other fundamental units, the progress of knowledge and of the practical applications of photometry required agreement on a more solid and universal unit of measurement for light. Therefore, in 1948, it was decided to use as a reference the light associated with the thermal radiation of a black body, which we discussed in the chapter devoted to the kilogram. It is a radiation whose characteristics are well known, which could be emitted by a very hot metal and, therefore, in a reproducible manner. Specifically, the chosen radiant material was platinum at its melting point at standard atmospheric pressure, equal to 1,768°C.

As in the case with the bar that defined the meter or the prototype kilogram, the reference for the candela was also a human artifact, and not even one that was easily reproduced: a piece of incandescent platinum. Thus, in 1979, thanks to the availability of increasingly more refined sources and detectors of luminosity, it was decided to liberate light from the artifact and to adopt a new definition. First, let me write it out for you, though I know it will appear cryptic. Afterward, we will try to clarify it. The definition reads: "The candela (cd) is the luminous intensity, in a given direction, of a source that emits monochromatic radiation of frequency 540,000 billion hertz and that has a radiant intensity in that direction of 1/683 watt per steradian."

I know, it's almost incomprehensible. Let's try to untangle the knots. First of all, let's start with this "source that emits monochromatic radiation of frequency 540,000 billion hertz." Put simply, it is a source that emits green light. Electromagnetic waves with a frequency of 540,000 billion hertz correspond, in fact, to a tone of green. This color was chosen for a reason: it is the frequency to which the human eye is most sensitive in daylight conditions. Next, the number 683. Admittedly, this seems like a random number. And, to tell the truth, to some extent, it actually is. Or rather it was chosen precisely because the luminous intensity of the green source of reference coincides as much as possible with the intensity of an actual candle. In this way, the new candela coincides with its historical definition from 1875.

The new approach of the international system of measurement, in which the fundamental units are all based on universal constants, has had relatively little influence on the definition of the candela. In substance, the chosen constant or reference is the luminous efficiency of the above-mentioned green-light source, and it was fixed as $K_{CD} = 683$ candelas per steradian per watt, where the watt—a secondary unit—is defined by way of the fundamental units of meter, second, and kilogram and the respective universal constants associated with them. Unlike the other universal constants, such as the speed of light in a vacuum c or the electron charge e, which are properties of the universe, K_{CD} is a very human constant. Martians coming down to Earth would have no problem agreeing on the value of c or e, but if they had a vision system different from ours they might consider the choice of the value of K_{CD} a highly arbitrary one, which was made merely for our convenience.

With the Utmost Satisfaction

Believing in science, apart from producing individual benefits, very often helps us to be active citizens and enlightened legislators. Grand Duke Leopold II, in addition to having confidence in preventive medicine, was also an innovator in criminal law. He was responsible for the so-called Leopoldina,  or Leopold Criminal Code, announced on November 30, 1786, which incorporated the ideas of Cesare Beccaria's juridical enlightenment and significantly reformed the criminal law framework of the Grand Duchy of Tuscany, making it the first modern state to abolish the barbarous institution of the death penalty. "With the utmost satisfaction of our paternal heart, we have finally recognized that the mitigation of penalties, conjoined with the most exact vigilance for the prevention of criminal acts, the swift expedition of trials, and the promptness and certainty of the punishment for true delinquents, rather than increasing the number of crimes, has considerably diminished the most common and made the atrocious almost unheard-of." Sound current?

Epilogue
Measures for Measure

We have come to the end of our journey of discovery of the seven measures of the world.

The international system of units of measurement is a powerful and universal tool for understanding nature, the world, and ourselves. Thanks to decades of work, metrology has finally arrived at a system of measurement that no longer depends on human experience but rather rests on unchangeable natural properties. If by some quirk of fate humans were to vanish from the Earth, together with all measuring tools that we have made—meters, balances, clocks—a new alien population who came to colonize the planet could reconstruct our system of measurement as is, because the speed of light and Planck's constant will never change. The system of measurement remains, however, a tool, and as such its primary added value lies in the mastery of those who use it. A chisel is a simple piece of iron as long as it sits on the table, but it becomes a tool that frees *David* from the marble when it is the hands of Michelangelo. In the same way, a set of electric and light measurements can be nothing more than a dry series of numbers, or they can become some of the experimental foundations of quantum mechanics, when interpreted by Albert Einstein in his analysis of the photoelectric effect.

Measuring is fundamental for our lives, for our well-being, and for the progress of human knowledge, but measurements must be made and used well. Now that we have discovered the beauty of that intellectual construction called a system of measures, we must recall the care and attention that must be put into its use, especially when it is being used for making collective decisions. By limiting ourselves to an incomplete set of measurements for describing or interpreting an event or a system we risk losing sight of complexity and many precious elements. A measuring process voluntarily limited to a subset of the quantities that describe a phenomenon can turn an impartial instrument of knowledge into a creator of distortions. The incredible leap forward of CT (computed tomography), which we encountered at the beginning of our journey, lies precisely in its having replaced the single point of view of traditional radiography with a multitude of observations from various angles.

If we want to describe the environmental qualities of electric cars, for example, it is not sufficient to limit ourselves to measuring the reduction of CO_2 emissions coming from each vehicle. We also need to consider where the electricity that powers those cars comes from and how much CO_2 was released in other places in order to produce it—places that may be quite far from where the energy is used. Otherwise, we focus on the clean air where electric cars are plentiful and believe that we have solved the problem, but we forget that if the electricity comes mainly from fossil fuels, as happens today, we are simply shifting the pollution from one place to another.

If we measure the success of a health care system only by the quantity of services it performs, without asking ourselves if sufficient resources are also being invested in the quality of those services and without determining whether the primary consideration is profit or the patient, we risk making wrong and discriminatory choices. If we base our evaluation of science more on the number of publications than on the quality and the impact of the scholarship, research has no future. If we evaluate people solely on the basis of aseptic, preestablished performance parameters, we lose part of our humanity and, incidentally, render less productive the environment in which those people work. If we renounce complexity and simplify all our evaluations with classifications, boxes, and divisions, we restrict ourselves to incomplete measurements that will inspire hypersimplified decisions and policies, incapable of constructing a future worthy of ourselves.

Measures are a precious tool, but they must be interpreted by humans, using science and its method. Conscious that measurement is a fundamental element of understanding, scientists have codified the measuring process and made it universal so that its results can be shared and verified at any moment and serve as the basis for theory. The analysis and choice of which quantity to measure are fundamental aspects of experimental practice, with the objective of describing as broadly as possible the system under examination and taking into account all potential points of view, whether real or imaginary. The discussion of experimental findings must be critical, and their reproducibility is fundamental to drawing solid conclusions.

Science is not an automatic dispenser of certainties, from which anyone can take what they need. On the contrary, scientific discoveries are the fruit of doubts and errors, which, for researchers, are not a reason for shame but rather a powerful instrument of knowledge. And they make science more human. Indeed, errors and doubts are just as fundamental for life as they are for research. Gianni Rodari, the great Italian educator and author of children's books, puts it this way in his *Il libro degli errori* (The Book of Errors), published by Einaudi in 1964: "Errors are necessary, useful as bread and often beautiful; for example, the tower of Pisa." The history of science is there to teach us that.

Great scientists like William Thomson, Baron Kelvin, Albert Einstein, and Enrico Fermi made mistakes. Kelvin erred in assessing the age of the Earth. Fermi thought he had found transuranic elements and did not realize that what he was observing was instead the fission of the uranium nucleus. Einstein introduced the cosmological constant to make relativity compatible with a universe that he mistakenly believed was static. All these errors, however, were fruitful. Kelvin's work, though it led to a mistaken result, nevertheless succeeded in transforming the study of the age of the Earth into a new science, which would soon determine the right result of 4.5 billion years. Fermi's conclusions on the presumed transuranic elements were the stimulus for the discovery by Lise Meitner, Otto Hahn, and Fritz Strassmann, made shortly thereafter, of uranium fission. Hahn himself admitted that he, Meitner, and Strassmann would never have been interested in uranium if it had not been for Fermi. Einstein's cosmological constant was actually an ingenious intuition, though at the time he arrived at it by erroneous hypotheses. Decades later, in fact, it would be rediscovered by astrophysicists to explain that the universe expands at nonconstant velocity.

These errors, like many others in the history of science, were generative, and they catalyzed turning points for scientific thought. Because, for every finding reached, for every measurement made, science does not stop. On the contrary, it poses ever more questions. The enthusiasm for a discovery is ephemeral, while doubt is a scientist's lifelong companion. Doubt that—to quote the Nobel Prize winner Richard Feynman—"is not to be feared" but to be "welcomed as the possibility of a new potential for human beings." For science, to doubt means to be free from reverential awe of pre-established authorities and ideas, because science is democratic and "one person, one vote" really does hold true, provided everyone has had the opportunity to bear the burden and the beauty of study. This is an opportunity that has to be given to everyone. This freedom allows us to take new paths, measure unexplored quantities, propose disruptive and revolutionary visions. A way of thinking that can also be applied outside of science, encouraging us not to wallow in a vision of progress and well-being that is purely financial and economistic.

If we put aside the simplistic narratives of much of our politics and mass media as well as the ex cathedra pronouncements that sometimes come from academia, the alliance between science and society can help make the ineluctable infinite complexity of the world that surrounds us accessible and manageable for everyone, without fear. Abandoning a forcedly coherent narrative—whether it be straightforward or esoteric—involves setting out on a narrower and steeper road, but one that can lead to messages whose content and value are much more elevated.

In the choice of tools with which to measure the world, humanity has entrusted itself to nature. We now have to entrust ourselves to the intelligence of individuals and communities so that those tools can permit us to invent the new measure for a sustainable relationship with nature and a well-being that is truly collective and universal.

Acknowledgments

In this book, you have encountered some very big numbers, such as Avogadro's number or the distance expressed in kilometers between the star Proxima Centauri and Earth. Even they are too small, however, to represent the amount of acknowledgments I owe to the many people who have helped me on this journey.

I begin with him without whom this book would simply not have come to be: Alessandro Marzo Magno, historian, excellent writer, and lifelong friend. I owe him my introduction to my Italian publisher Laterza, his encouragement to undertake a new project and return to book writing, as well as his attentive reading of the manuscript and his many valuable suggestions.

I asked many people for opinions; they all responded generously.

Giovanni Busetto, Alessandro De Angelis—Galileo scholar and popularizer—and Leonardo Giudicotti, my colleagues in the Department of Physics and Astronomy at the University of Padua, read the manuscript with care and gave me very useful advice.

I had the good fortune to meet Mauro Sambi, Professor of General and Inorganic Chemistry at our university, during the months in which I was writing. I am grateful to him for his meticulous review of the chapter on the mole, for the correction of my inaccuracies, and, above all, for having generously shared his time and reflections.

The revision of an expert on the subject such as Marco Pisani, physicist at the National Institute of Metrological Research in Turin, was crucial in improving various passages of the book.

To Maestro Federico Maria Sardelli goes my gratitude for having examined—and corrected—my improbable incursions into music and for a pleasurable explanation on the theme of musical tempi and the relationship between composer and performer.

I also owe a debt of thanks to Marina Santi, who reminded me in this period that not everything in life can or should be measured; to Alessandra Viola for her helpful advice; to Lia Di Trapani, Agnese Gualdrini, and everyone at the publishing house Laterza, who followed me and helped me in this adventure; to my colleagues, women and men, who, day after day, make the University of Padua, the RFX Consortium, and the DTT facility places of research and study in which I have learned so much and to which I am deeply indebted.

"Grazie" to Gregory Conti, who translated this book and did a great job, and a special thanks to Jean Thomson Black, from Yale University Press, who gave me the opportunity to be part of the prestigious YUP community, as well as to her editorial assistant, Elizabeth Sylvia, and manuscript editor Laura Jones Dooley for their fine work on the manuscript.

My heartfelt thanks go to all those who, over the years and in various ways, have stood by me in life, even when I didn't know how to recognize or appreciate their support. A special thanks to Annamaria and Carlo for everything they have taught me, in this period, too. And the most grateful thoughts to Andrea, my most important and precious reader.

All these people, and many others, have sustained and helped me with generosity. The errors and inaccuracies that remain in the book are exclusively my responsibility.

Suggestions for Further Reading

Without pretending to provide an exhaustive bibliography—the subject is indeed extremely vast—I suggest the following for those who might wish to undertake further investigation.

For a more in-depth view of how physicists use units of measurement, *The Feynman Lectures on Physics* is one of the best courses in physics. The approach is university level, but many parts are also accessible to nonspecialists. It is available online at www.feynmanlectures.caltech.edu. Also by Richard Feynman, and certainly a less academic work, *Six Easy Pieces: Essentials of Physics Explained by Its Most Brilliant Teacher*, 4th ed. (Basic Books, 2011), is well worth reading.

Revolutions in Twentieth-Century Physics (Cambridge University Press, 2012) and *Introduction to Quantum Mechanics*, 3rd ed. (Cambridge University Press, 2018), both by David J. Griffiths, are two specialized texts of great clarity. Equally clear despite its being a manual for specialists is *Quantum Physics: Of Atoms, Molecules, Solids, Nuclei, and Particles*, 2nd ed., by R. Eisenberg and R. Resnick (John Wiley and Sons, 1985).

Relativity: The Special and the General Theory (Methuen, 1920 and 1954) is a marvelous book, written by Einstein himself in 1916 as an introduction to the theory of relativity for the general reader. Along the same lines, a very complete book for general readers is *Einstein: The Life and Times* (William Morrow, 2007), by Ronald W. Clark.

On the theme of the history of measuring processes, Robert P. Crease is the author of the enjoyable *World in the Balance: The Historic Quest for an Absolute System of Measurement* (W. W. Norton, 2011). A very interesting article on the history of the concept of temperature is Martin K. Barnett, "The Development of Thermometry and the Temperature Concept," *Osiris* 12 (1956) and available online at jstor.org.

The Measure of All Things: The Seven-Year Odyssey and Hidden Error That Transformed the World, by Ken Alder (Free Press, 2002), is an interesting book on the story of Delambre and Méchain's adventurous measurement of the portion of the meridian from Dunkirk to Barcelona to determine the length of the meter. Chad Orzel's *Brief History of Timekeeping* (Oneworld, 2022) provides an interesting reading on the history of time measurements. Dava Sobel's *Longitude* (Walker, 1995) narrates how the challenge of measuring longitude at sea—a key problem for navigation—was solved in the eighteenth century.

Galileo Galilei is the father of the scientific method, based on experiments and measurements. *Galileo Galilei's "Two New Sciences,"* by Alessandro De Angelis (Springer, 2021), is an interesting book, aimed at making accessible to the general public Galileo's classic *Discourses and Mathematical Demonstrations Relating to Two New Sciences*, published in 1638.

In this book, I have used a few examples taken from my field of research, thermonuclear fusion. Readers who would like to have a historical overview of a field of physics that I find particularly fascinating and that boasts very elegant and complex high-tech measuring systems may wish to read *Nuclear Fusion: Half a Century of Magnetic Confinement Fusion Research*, by C. M. Braams and P. E. Stott (CRC, 2002).

A wide repertory of studies on units of measurement is available on the web. I would suggest the sites of the National Institute of Metrological Research in Turin, the American National Institute of Standards and Technology (NIST), and the Measurement Standards Laboratory of New Zealand, which has a curious interactive map that shows how measures and measurements are omnipresent in our daily lives. Two papers, Z. J. Jabbour and S. L. Yaniv, "The Kilogram and Measurements of Mass and Force," *Journal of Research of the National Institute of Standards and Technology* 106 (2001): 25–46, and R. S. Davis, "Recalibration of the U.S. National Prototype Kilogram," *Journal of Research of the National Bureau of Standards* 90 (1985): 263–83, contain useful information on the International Prototype Kilogram (IPK) replicas in the United States.

Also on the web, at the site www.nobel.se, readers can consult the keynote addresses delivered by Nobel Prize winners in the various disciplines, an interesting excursus among the pillars of modern physics.

A source of numerous technical details is the site of the Bureau International des Poids et Mesures (BIPM), www.bipm.org/en. Also technical, but interesting for specialists in the field, is the article published by D. B. Newall et al. in *Metrologia* (2018), entitled "The CODATA 2017 Values of h, e, k, and N_A for the Revision of the SI."

For readers of Italian, Alessandro Marzo Magno's *L'invenzione dei soldi: Quando la finanza parlava italiano* (The Invention of Money: When Finance Spoke Italian; Garzanti, 2013) provides a detailed account of the relationships between units of measurement for weight and currency names. Another work of history, *I diciotto anni migliori della mia vita* (The Best Eighteen Years of My Life), by Alessandro De Angelis (Castelvecchi, 2021), is an enjoyable account of the Paduan years of Galileo Galilei. *Il segreto degli elementi: Mendeleev e l'invenzione del Sistema Periodico* (The Secret of the Elements: Mendeleev and the Invention of the Periodic Table), by Marco Ciardi (Hoepli, 2019), reconstructs the history of the scientists and discoveries that led to the invention of the periodic table.

Enrico Fermi's eyewitness statement of the Trinity test is held at the U.S. National Archives, RG 227, OSRD-S1 Committee, box 82, folder 6 "Trinity," and reproduced online at www.dannen.com/decision/fermi.html. His letter to Edoardo Amaldi is held in the library of the Physics Department, Sapienza University of Rome, Edoardo Amaldi Collection, subcollection Amaldi archive heirs.

Finally, two books that, in my modest opinion, should not be missing from our personal libraries: *Il libro degli errori* (The Book of Errors), by Gianni Rodari (Einaudi, 2011), and *Il sistema periodico* (The Periodic Table), by Primo Levi (Einaudi, 1975).

Index

Index

Index

Index

Mussolini, Benito, 74, 75, 166–67, 167
mutual assured destruction, 143

nanoseconds, 47
Napoleon (Bonaparte), 25, 131–32
NASA (National Aeronautics and Space Administration), 66, 94
National Institute of Standards and Technology (NIST), 58
Natta, Giulio, 157
navigation, 55, 138, 194
Nazism, 32, 73, 76, 77, 78, 169
Neckam, Alexander, 139
neutrons, 135, 141
Newton, Isaac, 33, 59–60, 67, 93, 116; *Principia*, 95
newtons, 140–41
Newton's equation, 94
Newton's laws, 97, 98, 145
Ninth Symphony (Beethoven), 53
NIST (National Institute of Standards and Technology), 58
Noah's ark, 19–20
Nobel, Alfred, 99
Nobel Prize: in Chemistry, 76, 90, 157; in Medicine, 2; in Physics, 10, 16, 17, 28, 48, 74, 75, 78, 85, 91, 93, 96, 107, 188
Noctes Atticae (Attic Nights), 50
NSTX, 146
nuclear arms, 143. *See also* atomic bombs
nuclear energy, 74, 125–26, 143, 158, 176–77
nuclear fission, 125; chain reaction, 74; power stations, 103, 126
nuclear fusion, 83, 103–4, 126, 143, 146, 147–48
nuclear medicine, 158

nuclear power, 103–4, 126
nuclear reactors, 146–47, 176
nuclear waste, 126
nucleus (of an atom), 135
numbers, 163–66; Avogadro's number, 163, 165–66

Operation Valkyrie, 77
Oppenheimer, Robert, 74, 103
optics, 30
Ørsted, Hans Christian, 137, 138, 140
ounces, 81
oxidation, 159
oxygen: atoms of, 164, 165; and photosynthesis, 175–176; in water, 164–165

paces (*passuum*), 4, 21
Palatine Gallery of the Pitti Palace, 173
Papen, Franz von, 77
Parks, Rosa, 16–17
Pasteur, Louis, 108
pendulums, 51–52
pennies, 81
People's Republic of China. *See* China
periodic table, 155–56
Periodic Table, The (Levi), 154, 157
Persistence of Memory (Dalí), 60–61
Perzy, Erwin, 46
Philo of Byzantium, 111
photocatalytic processes, 161
photoelectric effect, 28, 91–92, 186
photometry, 180, 182
photons, 92
photosynthesis, 175–76, 179
physics: applied, 128; astro-, 188; atomic, 155; and atomic power, 27, 103; classical, 34, 88–89, 91, 93, 95, 97; electricity and, 135–36, 137; experimental, 134; experiments in, 34;